WILLIAM H. SHEPHERD PE, INC.
6877 N. HIGH STREET, SUITE 312
WORTHINGTON, OHIO 43085

Grain Storage

GRAIN STORAGE

The Role of Fungi in Quality Loss

Clyde M. Christensen and Henry H. Kaufmann

UNIVERSITY OF MINNESOTA PRESS   MINNEAPOLIS

Library of Congress Catalog Card Number: 70-76174

PREFACE

Grains and seeds are both exceedingly durable and highly perishable. If they are harvested sound and are subsequently kept at low moisture content and low temperature they may retain their original processing quality, and even their original germinability, for years or decades. But they are also susceptible to invasion and damage by insects, mites, and fungi, and if they are stored under conditions that promote the development of any of these deteriorating organisms, extensive spoilage may occur within a few days to a few weeks. Stored grains and seeds, like so many other stored products, are also subject to soiling, consumption, and contamination by rodents and birds. Good storage practices aim to maintain conditions in the grain that will preserve the marketing and processing qualities of the grain at as high a level as possible, and basically this means elimination of damage caused by insects, mites, and fungi, plus protection from rodents and birds.

Much basic and applied research has been devoted to insects of stored products, a lesser amount to mites. The means to control these pests are known, and in many countries they are being very effectively applied, as is indicated by the low degree of insect infestation in grains stored in commercial warehouses in the United States, in Canada, and in the developed countries in general.

The same is not true of fungi, since only recently have fungi been recognized as a major cause of spoilage in stored grains and seeds. Even today many of those who deal with grains, from warehousemen to those in the upper echelons of management, are unaware that storage fungi may play a decisive role in their operations, and in their profit and loss. Until fungi were recognized as a major factor in loss of quality of stored grains, and until the conditions that permitted or promoted damaging development of storage fungi were known, rational and practical methods for their control could hardly be developed. The revolution that has occurred in grain storage practices since about 1950 has been based in great part upon knowledge acquired from research on storage fungi. The now widespread aeration of bulk-stored grain with air of low temperature, for example, which is one of the important techniques in this revolution, could have been developed without anyone's knowing what was being controlled, but it developed much faster, more effectively, and with much less waste effort because of the information available about the nature and habits of storage fungi.

Among the first to recognize that fungi might be involved in the deterioration of stored grains and seeds were Drs. P. E. Ramstad and W. F. Geddes in the Department of Agricultural Biochemistry of the University of Minnesota. Their pioneering work of the late 1930's was continued by M. Milner and by others in that department, under the direction of Dr. Geddes. In the early 1940's work on this problem was undertaken by one of the authors (C. M. C.) and his students in the Department of Plant Pathology of the University of Minnesota, at first in cooperation with the Department of Agricultural Biochemistry but later carried on independently. In 1952 Cargill, Inc., instituted a fellowship in the Department of Plant Pathology, devoted to the study of the cause and control of spoilage of stored grains by fungi, and this fellowship has been continued down to the present. Also in 1952 Cargill, Inc., established its own Grain Research Laboratory, under the direction of the joint author (H. H. K.) for the study of some of the basic and practical aspects of maintenance of quality in stored grains. Collaboration in this work between Cargill, Inc., and the Department of Plant Pathology, and between the two authors, has been close and continuous for more than fifteen years, and by far the major portion of the research efforts of both of us during this time has been devoted to the problems of grain storage. Although much of our laboratory and field or warehouse work has dealt with grains produced and stored in the United States, both of us have at various times had considerable experience with grain storage problems in other countries.

Many investigators throughout the world have devoted a good deal of study to the solution of one or another of the problems involved in the maintenance of quality in stored grains and seeds. Most of the results of such research have been published in technical journals, but the information never has been presented as a coherent whole in a fashion understandable by the nonspecialist. It was thought that a summary of these research results might be of interest and value to all those who buy, sell, store, and process grains and seeds, and this book is an attempt to present such a summary.

Chapter 9, on insects, mites, and rodents, is somewhat outside the province of the main theme of the book — the nature and control of damage by fungi. However, some kinds of insects and mites that infest stored products are very closely associated with storage fungi, and the conditions that limit the activities of insects and mites also limit, in general, the activities of storage fungi. For these reasons the inclusion of the material in this chapter seemed justified.

Naturally we are indebted to many of our co-workers for support, criticism, and advice in our research work over the years, and in the preparation of the manuscript for this book. To all of them we are grateful. Preparation of the manuscript was facilitated by a grant from the Rockefeller Foundation, for which we express our sincere thanks.

C. M. C.
H. H. K.

October 1968

TABLE OF CONTENTS

Illustrations between pages 70 and 71

Grain Storage

1 THE PROBLEM OF LOSSES IN STORED GRAINS

Grains and seeds constitute a major source of foods and of raw materials for industrial processing. In 1966 there were produced in the world about 10,100,000,000 bushels of wheat and 377,450,000,000 pounds of rough rice (rice grains with the hulls on), equivalent to about 260,000,000,000 pounds of milled rice (rice seeds divested of the hulls), nearly all of which was used for human food (6). In the same year the world production of corn (maize) was about 8,500,000,000 bushels, used for food for man and domestic animals, for the manufacture of industrial and potable alcohol, and for processing for starch, oil, and other materials. In some of the Latin American countries corn has long formed the main portion of the diet of man, as it does in some countries of Africa at the present time, where its importance is increasing. In one or another portion of the world other grains and seeds such as oats, barley, millet, sorghum, buckwheat, beans, soybeans, peas, or various combinations of them are vital ingredients of the diet. All these kinds of grains and seeds are subject to losses during storage.

MAGNITUDE OF LOSSES

Officials in the Food and Agriculture Organization (FAO) of the United Nations, who probably are in as good a position as anyone to give a fair appraisal, have estimated that 5% of all food grains harvested are lost before consumption. The magnitude of losses will vary from country to country and from year to year, of course, but in India, in portions of Africa, and in some South American countries it is estimated that 30% of the annual harvest is lost. If the FAO estimate of 5% over-all loss is approximately correct, about 50 billion pounds of the 1966 crop of wheat and rice were lost, or enough to give 100 pounds more per year to each of 500,000,000 people. The losses are likely to be higher in the countries that can least afford them — in part because of climates favorable to deterioration of all kinds of stored products, but in part also because of lack of knowledge and facilities necessary to reduce or prevent such losses. Many of these losses are needless, a result of ignorance or carelessness, or both. In some of the countries where heavy losses in storage are the rule, great efforts are being made to increase the production of edible grains and seeds, but little or no effort is made to improve storage facilities, or even to study the problem. The people need food, not production statistics, and a crop is not food until it is eaten. A program to reduce storage losses probably could result in a 10%–20% increase in available food now in some of the developing countries, and might also assure that whatever increases occur in production in the future would be used for the nourishment of people, not for feeding pests.

CAUSES OF LOSS

The principal causes of loss in quantity and quality of stored grains and seeds are rodents, insects, mites, and fungi. As indicated in the Preface, this book is concerned almost exclusively with the problems caused by fungi, for the reasons there given.

NATURE OF LOSSES

The major types of losses caused by fungi growing in stored grains are these: (1) decrease in germinability; (2) discoloration of part (usually the germ or embryo) or all of the seed or kernel; (3) heating and mustiness; (4) various biochemical changes; (5) production of toxins that if consumed may be injurious to man and to domestic animals; (6) loss in weight. A good deal of research has been devoted to some of these changes, since they affect the quality of grains for most of the uses to which grains are put, and in the following chapters the results of this research are summarized. In a given case of spoilage one or more of these deleterious changes may have occurred, depending on the stage to which the spoilage has progressed. Also in any given case of spoilage caused by storage fungi the first three of the changes listed above occur in the order given, that is, decrease in germinability, followed by discoloration of part or all of the seed, followed by heating and mustiness. Some biochemical changes and some loss in weight will have accompanied the other changes from the beginning.

All these changes, including production of toxins, may occur without the fungi responsible for them becoming visible to the naked eye. It is in part the inconspicuous nature of these fungi that is responsible for their belated recognition as major causes of damage in stored grains. Practical men, and even some trained researchers with little training in microbiology and none in mycology, were reluctant to believe that organisms they could not see were capable of causing the drastic and sometimes dramatic changes that were observed. However, this inconspicuous nature is not something peculiar to storage fungi growing in stored grains — it is the usual thing in plant tissues invaded and decayed by fungi. Wood, for example, may be decayed so thoroughly that it can be crumbled into powder, and yet the fungus that causes the decay is not visible to the naked eye — and sometimes the fungus is not visible with the aid of a microscope. Similarly, in the tissues of potatoes, apples, or other plants or plant parts decayed by fungi, the causal fungi usually are not visible to the naked eye, and often they are not visible with the aid of a microscope. When the decay has progressed to its final stage, and the invaded tissues have turned brown or black and have been largely consumed, the fungi may or may not produce a crop of spores visible to the naked eye. Grain in the final stages of decay by storage fungi may be caked and bound together by fungus mycelium, and may be powdery with spores and have a strong musty odor — but this is the final stage of decay, not the beginning stages. Many common storage problems, many cases of severe losses in stored grains, are the result of invasion by fungi that one cannot see. To better illustrate the nature of the losses that storage fungi may cause, a few concrete cases will be described in detail.

Corn or Maize. The maintenance of quality in corn or maize during storage has long presented serious problems, in part because of the tremendous quantity handled and the diversity of uses to which the grain is put, but in part also because of the different moisture contents permitted in the different grades. In normal years, by far the greatest amount of corn in national and international commerce is grade No. 2 (or, in those corn-producing and exporting countries that do not have formal grade designations, corn with characteristics comparable to those of grade No. 2) in which a moisture content of 15.5%, wet weight basis,

5

is permitted. Most of the grade No. 1 corn that gets into trade channels is likely to be mixed with corn of grade No. 3 or No. 4 to bring the latter up to the specifications of grade No. 2. The implication is that corn with a moisture content of 15.5% is safe for storage, and for a long time practical grain men believed that it was. Actually it may or may not be safe for storage, depending on the temperature of the grain, the length of time it is to be stored, and the degree to which it has been invaded by storage fungi, to name the major factors that affect storability. The use to which it is to be put will also influence a buyer's judgment on what constitutes spoilage: a higher quality will be demanded for cornflakes or cornmeal than for pelleted feed. Among different lots of No. 2 corn there may be great differences in the degree of invasion by storage fungi and in the rate at which deterioration will occur if the grain is stored under conditions that permit the fungi to grow. It is commonly stated that with the advent in the corn belt of the United States in the mid-1960's of the picker-sheller, which requires a moisture content in the corn of about 23% for most efficient operation, the problems associated with the storage of moist corn greatly increased, and will continue to increase. However, spoilage of moist corn was a serious problem more than sixty years ago, as a few extracts from the literature of that time will show.

Shanahan *et al.* (5) in 1910 stated: "For several years [in the early 1900's] an increasingly large number of more or less forcible and persistent representations were made to the Secretary of Agriculture and other officers of the Federal Government, to the effect that much of the grain, and especially the corn, that was being exported from the United States was not being delivered abroad in a satisfactory condition, that it was not of the quality represented by the inspection certificates accompanying the shipments, and that material injury was in consequence being done to the export grain trade of the United States."

Even at that time the export trade in corn was considerable: from 1898 to 1908 about 100,000,000 bushels annually, mostly to western European countries. In 1966 the export of corn from the United States exceeded 500,000,000 bushels.

Shanahan and his co-workers began in 1906 to investigate the condition of exported corn, and continued the investigation for several years, eventually getting samples from 141 corn-laden ships at 34 European ports. These ships carried a total of 15,077,987 bushels of corn, all of which had been certified as No. 2 or equivalent grade, such as "sail grade" or "prime sail grade" (this was before federal grain standards were established), of which 1,911,374 bushels, or 12.7%, arrived in a heating or hot condition, some portions damaged so much as to be unfit for feed. In 8 of the 141 ships, upon arrival at European ports, 100% of the corn was heating or hot. The moisture content of the grain ranged from a low of 13.8% to a high of 28% when it was unloaded. When loaded at American ports, 43 cargoes had moisture contents of 14.0%–16.0%, and 37.2% of these were heating when they arrived in Europe. Sixty-four cargoes had moisture contents of 16.1%–18.0% when loaded, and 73.4% of these were heating or hot when they arrived in Europe. This was corn being sold by the hardheaded Yankee, and bought by the astute and sophisticated European — both close-fisted, practical, down-to-earth businessmen with a sharp eye for a dollar, franc, mark, or kroner. It sounds fantastic, not only because the traffic in spoiled corn was so huge, but also because the same traffic in spoiled corn continued unabated for years after the investigation by Shanahan and his coworkers.

These men were interested in determining the cause of the spoilage, so that rational steps could be taken toward its reduction. They studied the relationship to spoilage of such things as season of the year when grain was shipped, the location of the grain in the ship (whether near to or far from engine room or shaft alley bulkheads), length of voyage, temperature of the grain, and moisture content of the grain. All these things had been suggested as possible contributing factors. The grain men themselves heavily favored the theory that the season of the year the grain was shipped was the main factor. According to Shanahan *et al.*, "In the grain trade the 'germinating season,' so called, is understood to be a special season of the year during which grain is customarily planted in the ground. The limits of this season are not very clearly defined, but it is generally understood to extend from about the middle of March to the middle of June. It is generally believed that there is a natural and inherent tendency in grain to germinate during that season, and that the heating of grain in storage and in transit during those months is due primarily to this factor."

Shanahan did not accept this explanation, which puts the blame on the corn, although one still hears it advanced by practical, serious, experienced grain men in charge of millions of bushels of stored corn. Shanahan *et al.* stated: "These investigations have led to the conclusion, which is believed to have been clearly demonstrated in the tables and diagrams, that the moisture content of the corn, and of other grains as well, is the primary factor determining their capacity to carry safely in ocean transit without deterioration, and the importance of this factor has been emphasized throughout the work. Corn in which the moisture content is sufficiently low will carry safely under ordinary conditions of ocean transit for any reasonable length of time during any season of the year, no matter where it is stowed in the vessel, while corn containing a high moisture content is constantly in danger of heating at any time owing to a variety of contributing causes." These investigators do not appear to have been aware of the possible primary role of fungi in the deterioration process, but Black and Alsberg (2) were, because in their bulletin, also published in 1910, they stated: "The tests for micro-organisms and the tendency to become moldy involves the quantitative determination of the number of organisms in the suspicious sample compared with a sound sample." This sound and perceptive idea was disregarded for the better part of thirty-five years.

The information published by Shanahan *et al.* must have got around, because within a few years the bulletin was out of print, but if it made any impression on the practical men in the grain trade it certainly did not greatly alter their practices, as shown in another investigation undertaken by one of the authors, Boerner, the results of which were published in 1919 (3). He and his coworkers took corn samples from 41 holds in 9 ships and tested the grain for various characteristics. Only 12 of the 41 holds contained corn with a moisture content *below* 18% when loaded, and only 1 of the 41 had corn whose moisture content was below 15.5%; the grain in this single exception had a moisture content of 14.9%. The grain in 31 of the 41 holds was heat damaged upon arrival at the European ports, and in 30 of these the temperature of the grain when unloaded ranged from 119° to 155° F. The grain in some of the holds was so caked and solidified by fungus mycelium that it had to be chopped out.

Probably the only reason the grain investigated by these men ever got from the corn belt of the United States to the loading ports without extensive spoilage is that it was kept at a temperature below 10° C. (50° F.) from shortly after harvest until it was loaded into the ships. It is highly likely that much of it, by that time, was so heavily invaded by storage fungi that it was on the verge of spoilage, and all that was needed was a slight rise in the temperature of the grain to allow the fungi to grow rapidly.

The problem of spoilage of grains in transit has been reduced, but not eliminated. In 1965 we received from grain importers in Amsterdam, Holland, corn shipped from Brazil — corn which had been heavily invaded, heated, and decayed by fungi, including *Aspergillus flavus* and others known to produce toxins. The importers stated that shiploads of corn that had spoiled in transit were "common." Also in 1965 we obtained 376 samples of corn being loaded into 36 ships at United States ports from New Orleans to Baltimore for export to Europe. None of these had moisture contents above 15.5%; most of them had moderately high germinability, moderate to low invasion by storage fungi, and moderate to low fat acidity values. According to our evaluations and actual storage tests of these samples, none of them was of poor storability. Unquestionably there has been great improvement in the quality of corn exported from the United States since the 1900–1915 era, but probably there is room for more improvement.

If corn spoils during ship travel on the high seas, the blame cannot always be placed on the shipper. In May 1962, a cargo of somewhat more than 100,000 bushels of No. 2 white corn, in bags, was loaded into the holds of a ship at New Orleans. The ship sailed from New Orleans on May 29, passed through the Gulf of Mexico, through the Panama Canal, and up the west coast to El Salvador, where it arrived on June 17, after a voyage of only eighteen days. On arrival, much of the corn was spoiled, and during its storage in a warehouse near the unloading docks the rest of it spoiled. The total loss was more than $200,000, and in the resulting lawsuit for the recovery of damages, the question at issue was whether the corn had been in bad condition and prone to spoil when loaded into the ship or whether improper care during the voyage had resulted in the loss.

Fortunately, a sample of the corn as loaded into the ship had been kept and was still available, a precaution that if more generally followed would establish almost without question the quality and condition of grain when loaded for export. This sample, presumably representative of the bulk as loaded, when tested two years later still germinated 80%, had less than 2% damaged kernels, and was free of storage fungi. Samples of the corn taken from bags as these were unloaded in El Salvador had 0% germination and from 20% to 40% damaged kernels, and were heavily invaded by a variety of storage fungi. From the *kinds* of storage fungi that grew from surface-disinfected kernels that were cultured, we knew that this corn had to have been exposed to a relative humidity of 85%–90% for at least two weeks at a moderately high temperature — for only under such conditions will those particular fungi invade corn.

The ship's log showed that the relative humidity of the air during the voyage ranged from about 85% to 90%, and the temperature was mostly between 80° and 85° F. The ship had no forced ventilation of the holds, but had kept the scoop ventilators open and turned into the wind throughout the voyage, except when rain was falling. So the corn was continually exposed to a blast of warm, moist air. We obtained samples of corn of the same class and grade and from the same location that the corn loaded into the ship had come from, and exposed samples in small bags to a relative humidity of 85%–90% and a temperature of 80°–85° F. for eighteen days; this corn was then found to be in just about the same condition as the corn in the ship when it arrived in El Salvador. It was invaded to about the same degree and by the same kinds of fungi as was the corn that arrived spoiled in El Salvador. This was convincing evidence, and the case settled in favor of the grain firm that had supplied the corn.

Wheat. Wheat is just as subject to spoilage as is corn, under conditions that permit storage fungi to grow vigorously, and periodically in the United States heavy losses occur, especially in quality. For example, wheat bought and stored as milling grade is, frequently during storage, reduced to feed grade. Such losses are likely to be widespread when wet weather prevails at harvest time over relatively large areas of the main wheat-growing regions. This may result in much wheat coming into storage with a moisture content too high for safe storage. Until 1964 the Official Grain Standards of the United States Department of Agriculture (USDA) permitted a moisture content of 14.0% in winter and white wheats, and 14.5% in hard red spring and durum wheats. The assumption was that the hard red spring and durum wheats were more resistant to deterioration in storage than the other classes of wheat, and so could be permitted to have 0.5% more moisture. Damaging invasion of wheat by storage fungi occurs first in the germs or embryos, and the embryos are separate and distinct from the endosperm of the seeds. Hardness or softness in wheat seeds refers to characteristics of the endosperm, and the embryo or germ has nothing to do with hardness or softness. The hard red spring and durum wheats were less damaged by storage fungi than other wheats, at moisture contents and temperatures that permitted development of storage fungi, because they were grown and stored farther north, where the temperatures at harvest and during subsequent storage were low. If samples of different classes of wheat, of equal soundness, are stored at, say, 16.0% moisture content and 25° C., all of them will be invaded by storage fungi at an approximately equal rate, and suffer about an equal amount of damage in a given time. This eventually was recognized, and in the 1964 revision of the Official Grain Standards of the United States (7), the upper limit of moisture content for all classes of wheat was set at 13.5%. From the standpoint of maintaining quality in storage this was a desirable change.

Soybeans. Soybeans have long had the reputation among elevator men of being "difficult to store." At moisture contents above about 11.0% indeed they may be difficult to store without loss of quality, because they will be invaded and spoiled by storage fungi. The moisture content limits for grades No. 1, 2, 3, and 4 of soybeans, as given in the Official Grain Standards of the United States, are 13.0%, 14.0%, 16.0%, and 18.0% respectively, wet weight basis. If soybeans are kept at the moisture content specified for grade No. 1, 13.0%, and at a temperature favorable for the growth of fungi, they will be invaded slowly by storage fungi, and if they are kept at a moisture content of 14.0% or above they will be invaded and damaged by storage fungi within a few weeks, unless the temperature is so low that fungi cannot grow rapidly. In other words, the moisture content limits specified for grades No. 1 and 2 soybeans are too high for safe storage for more than a short time, and so if practical grain men sometimes encounter trouble in storing soybeans, the fault is by no means theirs alone. If soybeans are kept at moisture contents and temperatures that will not permit storage fungi to grow, they are no more difficult to store than are any other kinds of seeds.

Flaxseed. This seed crop is produced in quantity in northwestern Minnesota and eastern North Dakota, and much of it is processed in the Twin Cities, only a few hundred miles away; the major product is linseed oil, used in paints and in other materials where quality is not so important as it is in foods. This combination of short haul from producer to processor, relatively low quality requirement, and the moderately cool, dry climate of the region where the seed is grown and processed, should minimize storage problems. Perhaps it does, but it by no means eliminates them.

The United States Grain Standards specify a moisture content no higher than 11.0% for grades No. 1 and 2 (the only numerical grades) of flaxseed, and by this is meant the average moisture content of a representative sample of the lot in question. As will be discussed in Chapter 3, this figure gives no indication of the *range* of moisture content in a given lot of grain or seed, and frequently leads to a false sense of security regarding the probable keeping qualities of a given lot. The 11.0% moisture content limit specified for grades No. 1 and 2 of flaxseed implies that this is a safe moisture content for storage without loss of quality. Safe for storage for how long and at what temperature? Flaxseed with 11% moisture content and at a temperature that permits fungi to grow will be slowly to rapidly invaded by storage fungi, the speed of invasion depending chiefly on the temperature. Normally flaxseed in North Dakota is harvested rather late in the fall, and so it goes into storage on farms at a moderately low temperature. Later it is transferred to country elevators, where it also remains at low temperatures, unless it is undergoing heating, which is one of the products of rapid growth of storage fungi. Ordinarily the storage fungi do not invade the flaxseed enough to cause any obvious mustiness within a few months after harvest, and unless the seeds are examined with the aid of a dissecting microscope and by someone familiar with storage fungi, or are cultured to detect storage fungi, they appear to be sound, whereas actually they may be on the verge of spoilage. Flaxseed is likely to be shipped to the processing plants in late fall, where it is stored in large bins until the oil is extracted, and in these bins it sometimes heats and spoils.

In the 1950's, before a USDA board that sets grain standards, hearings were held concerning the moisture content limit of flaxseed. The commercial processors of flaxseed wanted the moisture content limit lowered because, they contended, 11.0% moisture content led to heating and spoilage of flaxseed. Representatives of the flax growers of North Dakota were opposed to any such lowering of the moisture content limit (no producer of grains and seeds *ever* wants the moisture content limits reduced), and claimed that they never encountered any spoilage or heating in flaxseed stored with a moisture content of 11.0%. The discussion at times waxed warm, and naturally so, since at that time the water in flaxseed was worth about 10¢ per pound. Neither side realized that storage fungi were the primary cause of the trouble, and probably would not have believed it had they been told.

Both sides were right. The flax growers had no trouble with flaxseed at 11.0% moisture content because they seldom kept the seed long enough for obvious spoilage to occur — and to them obvious spoilage would mean grain so caked together with fungi that it could not be readily transferred from the bin to the truck that was to haul it to the country elevator. The beginning stages of spoilage would go unnoticed by them, and so a good deal of flaxseed arrived at the terminals and processing plant elevators just on the verge of spoilage, and some of it subsequently heated and spoiled. The processors were quite right in maintaining that 11.0% moisture content was too high for safe storage of flaxseed, since storage fungi will invade flaxseed with a moisture content of about 10.5% and above. Anyone who intends to store flaxseed for more than six months at a temperature above 20°–25° C. (68°–77° F.) should be certain that none of the flaxseed has a moisture content above about 10.0% — and this doesn't mean just that the average moisture content of a representative sample should be 10.0%, but that nowhere in the mass of bulk or lot should any of the seed have a moisture content above 10.0%. The moisture content limit of flaxseed, by the way, was not changed as a result of the hearing.

Sorghum. Storage fungi may also reduce germinability of seed intended for planting, although such losses are not common because this seed is expensive and usually is given special care. One case of interest may be cited. In 1965 about 93,000 bushels of sorghum seed, supposedly of certified seed grade, were stored in metal bins of about 2000 bushels capacity each, in Nebraska. It was worth about $8.00 per bushel. By the fall of 1966 the germination percentage had decreased to the point where the grain was fit only for feed, worth about $1.00 per bushel. It was claimed that the decrease in germinability was the result of fumigation with methyl bromide, which when used in sufficient concentration may be toxic to seeds, especially if they have a high moisture content. Tests of samples of these seeds from the surface and from a six-foot depth in different bins indicated that the loss of germinability probably was due to invasion of the seed by storage fungi. Some of this seed evidently was stored with a moisture content just below that needed for storage fungi to grow. In the winter of 1965–66, moisture increased in the center of the upper portion of the grain in these bins, probably because of slow circulation of air up through the center and down the outside. By spring the grain in the upper central portion of the bins was moist enough so that when the temperature rose in the spring, the fungi grew, invaded the seeds, and killed them.

We have not worked with fungus-free sorghum seeds, which are very difficult to obtain, but we have worked with fungus-free seeds of corn, which is not too distantly related to sorghum, and these will endure a surprisingly high temperature and moisture content for months without much loss in germinability, so long as they remain free of fungi. Effects of storage fungi on germinability of seeds are described in Chapter 5.

Cases similar to those described above could be cited almost without end, involving all kinds of grains and seeds as well as other plant products such as baled hay and baled cotton, and some animal products, such as baled wool, wherever in the world these products are stored, carried, or handled.

GRAIN GRADES AND STORABILITY

The United States Grain Standards Act, passed in 1916, established the administrative nucleus of the grain inspection service. Detailed specifications for different grades of corn were established July 1, 1916, and those for wheat went into force on July 1, 1917; specifications for other grains came later. These specifications are summarized in a booklet, *Official Grain Standards of the United States* (7). Some of the specifications have been revised occasionally, the last time in 1964. The specifications for wheat and corn, from the Official Grain Standards, are given in Tables 1 and 2.

The grain grades attempt to maintain certain standards for merchandising — for buying and selling. They do not evaluate storability, nor is this their function.

Mexico has grading rules of a sort, although these are not so formalized as those in the grain trade in the United States; essentially their specifications conform to those of grade No. 2 of corn, wheat, sorghum, and beans. More stringent and formal specifications have been suggested by Genel (4). So far as we are aware, in much of the Near East, Far East, Africa, and South America there are no generally adopted grades for edible grains and seeds, and no formal inspection as is practiced in the United States and Canada.

Table 1. Requirements in Grades 1 through 5 for All Classes of
Wheat except Mixed Wheat*

Requirements	1	2	3	4	5
Minimum test weight per bushel (in pounds)					
Hard red spring wheat	58.0	57.0	55.0	53.0	50.0
All other classes	60.0	58.0	56.0	54.0	51.0
Maximum limit of defects (in percentage)					
Heat-damaged kernels	0.1	0.2	0.5	1.0	3.0
Total damaged kernels	2.0	4.0	7.0	10.0	15.0
Foreign material	0.5	1.0	2.0	3.0	5.0
Shrunken and broken kernels	3.0	5.0	8.0	12.0	20.0
Total defects	3.0	5.0	8.0	12.0	20.0
Maximum limit of wheat of other classes (in percentage)					
Contrasting classes	0.5	1.0	2.0	10.0	10.0
Total wheat of other classes	3.0	5.0	10.0	10.0	10.0

Source: *Official Grain Standards of the United States* (7).
 * In addition to the five numerical classes, sample grade is identified: "Sample grade shall be wheat which does not meet the requirements for any of the grades from No. 1 to No. 5, inclusive; or which contains stones; or which is musty, or sour, or heating; or which has any commercially objectionable foreign odor except of smut or garlic; or which contains a quantity of smut so great that any one or more of the grade requirements cannot be applied accurately; or which is otherwise of distinctly low quality."

Table 2. Grade Requirements for the Classes Yellow Corn, White
Corn, and Mixed Corn*

Requirements	1	2	3	4	5
Minimum test weight per bushel (in pounds)	56	54	52	49	46
Maximum limits of defects (in percentage)					
Moisture	14.0	15.5	17.5	20.0	23.0
Broken corn and foreign material ..	2.0	3.0	4.0	5.0	7.0
Damaged kernels					
Total	3.0	5.0	7.0	10.0	15.0
Heat-damaged kernels	0.1	0.2	0.5	1.0	3.0

Source: *Official Grain Standards of the United States* (7).
 * "Sample grade shall be corn which does not meet the requirements for any of the grades from No. 1 to No. 5, inclusive; or which contains stones; or which is musty, or sour, or heating; or which has any commercially objectionable foreign odor; or which is otherwise of distinctly low quality."

Grain Storage

It is, of course, the privilege of a grain buyer to specify just about any characteristics that he desires in the grain he buys. If he wants corn with a moisture content specified for grade No. 2, but with the other characteristics of grade No. 3, he can get it, or if he wants corn with 14.0% moisture and no more than 2% damaged kernels, he can get it. Grain merchants are knowledgeable men, and many of the grain merchandising and processing firms in this and other countries have been in business for many years; presumably they are making a profit or they would not survive. If they buy grade No. 3 corn, with a top limit of 17.5% moisture, they know the approximate degree of risk they are taking. They balance this against the chance of profit, and make their decision on the basis of experience. Often a calculated risk is taken since, as stated by Bailey (1), "If we never have condition troubles, we are playing it too safe." Development of techniques to evaluate storability and to maintain quality in storage permits the grain merchant to calculate his risk much more exactly and rationally than he could twenty years ago.

2 CHARACTERISTICS OF FIELD AND STORAGE FUNGI

Work over the past thirty years has established beyond all doubt that some of the fungi that invade stored seeds have a tremendous influence on the grade, condition, and keeping quality or storability of grains. Acquaintance with these fungi and understanding of how and where and why they grow is desirable for those who deal with stored grains and seeds, because one of the requirements of good storage is the prevention of growth of these fungi. Those who would like more background information on fungi than can be given here may consult two books (10, 15) that serve to introduce fungi to the lay reader.

More than 150 species of fungi have been isolated from cereal seeds, and from a single gram of malting barley, about 25 kernels, we have isolated tens of thousands of colonies of filamentous fungi, hundreds of thousands of colonies of yeasts, and several million colonies of bacteria. Malting grade barley is about equivalent to seed grade — the best of the barley; and while the degree of invasion or contamination indicated by the figures cited may not be typical, it is not at all unusual. Such contamination or invasion of grains of all kinds is likely to be high in regions or seasons of wet harvest, and low in regions or seasons of dry harvest weather, and is likely to be higher in seeds of cereal plants, which are borne relatively exposed to the air, than in beans, peas, or soybeans, borne in pods. We divide the fungi that invade grains and seeds into two groups, primarily on the basis of their behavior: field fungi and storage fungi.

FIELD FUNGI

These fungi invade the kernels or seeds before harvest, while the plants are growing in the field, or after the grain is cut and swathed but before it is threshed. There are some exceptions to this, notably corn

17

stored on the cob in cribs and exposed to the weather; under such circumstances it may be invaded by field fungi, or the field fungi already present in it may continue to grow. The predominant field fungi differ somewhat according to the crop, the region or geographic location, and the weather, but in wheat, rice, barley, and oats, grown in much of the world, the major field fungi that invade the kernels are species of *Alternaria*, *Cladosporium*, *Helminthosporium*, and *Fusarium*.

Alternaria is common in many grains and seeds, especially the cereals, but it is not restricted to the cereal seeds. It is, for example, a predominant fungus in peanuts. We regularly isolate *Alternaria* from close to 100% of surface-disinfected wheat kernels after harvest. If, in fact, we obtain *Alternaria* from 90% or more of wheat kernels of a given lot that are surface disinfected and placed on an agar medium in culture dishes, but obtain no storage fungi, we consider this good evidence that the seed is newly harvested and that it has been stored under conditions that do not permit deterioration. If it had been stored at too high a moisture content, even for a few weeks, and were beginning to go out of condition, *Alternaria* would have begun to die out and storage fungi would have begun to increase. In other words, the relative proportion of kernels that yield *Alternaria* tells us something about the condition of storage of that lot of grain, information of value in judging storability of the seed, and information that we cannot always get from warehouse records.

Cladosporium is common in cereal seeds that have been exposed to moist weather during harvest, especially grains harvested with the hulls on, such as barley, oats, and rice. It may cause darkening of the invaded hulls, but has no known effect on storability.

Helminthosporium is common in many lots of cereal seeds especially if the weather just before harvest has been moist. It may cause discoloration of the seed, death of the germinating seed or young seedling or root rots and blights of the mature plant, but causes no loss in storage.

Fusarium also is common in freshly harvested cereal seeds. Some strains or species of it cause "scab" in barley, wheat, and corn; scabby grain may be toxic to animals, including man, and so is undesirable for food or feed. The subject of toxicity of grains invaded by fungi is taken up in Chapter 6, and will not be discussed further here. We have some evidence that infections by *Fusarium* too light to be detected by visual inspection (but detected by culturing surface-disinfected kernels in agar) may result in later death and discoloration of the germs of the stored grains. This should be investigated more fully, because if true, it means that field infections by *Fusarium*, even if too light to cause reduction in grade of the grain just after harvest, might be responsible for development of germ damage, and corresponding reduction of grade and price, later in storage. *Fusarium* dies relatively rapidly in grain stored at moisture contents about 12%–13% and with temperatures above 70° F., and after it has died there is no way to detect that it once was present.

With this possible exception, the damage that field fungi do to grain is done by the time the grain is harvested, or at least by the time the grain is dried to a moisture content below 20%–22%. All the field fungi require a high moisture content in seeds in order to grow — a moisture content in equilibrium with a relative humidity of 90% or more (in the starchy cereal seeds, 20%–21% moisture, wet weight basis). The field fungi may survive for years in dry grain (14), but die fairly rapidly in grains held at moisture contents in equilibrium with relative humidities above 70% — in the starchy cereal grains, this means moisture contents above about 14.0%. Recent evidence, in fact, indicates that with the right combination of moisture content, temperature, and time of storage it is possible to rid barley seeds completely of field fungi, with no invasion by storage fungi and little or no reduction in germination percentage of the seed or vigor of the seedlings (18). We do not yet know whether this might be of practical value.

In summary, field fungi may affect the appearance and quality of seeds and grains for almost all purposes for which seeds and grains are used. Usually, damage caused by field fungi is caused before harvest, can be detected by routine inspection, and does not continue to increase in storage. The field fungi present in seeds at harvest gradually die; the length of time they survive depends mainly on the moisture content and temperature of the stored seeds. Numbers and kinds of fungi in grains can be determined by culturing surface-disinfected kernels on agar, and data so obtained may aid greatly in judging the storage history and future storability of a given lot of grain.

NATURE OF STORAGE FUNGI

Kinds of fungi and general characteristics. The storage fungi comprise about a dozen species of *Aspergillus* (of which only about five are at all common), several species of *Penicillium* (which we lump together without trying to separate the different species, partly because they have much the same moisture requirements, partly because identification of species of *Penicillium* with any assurance is a task for a specialist), a single species of *Sporendonema*, and possibly a few species of yeasts. In practice one may culture 50–100 surface-disinfected kernels from each of hundreds or even thousands of lots of wheat and encounter few storage fungi other than *Aspergillus restrictus* and members of the *A. repens–A. amstelodami–A. ruber* group, all of which, for practical purposes, may be grouped together.

All these, as will be taken up in more detail below, have the ability to grow in grains and seeds whose moisture contents are in equilibrium with relative humidities of 70%–90%. Most of these fungi are common on a great variety of organic and inorganic materials, especially decaying vegetation, food products, fabrics and insulating materials made of plant fibers, paints and coatings, leather goods, and glues. They occur almost everywhere throughout most of the world, they are among the most successful and abundant of living things, and they almost inevitably contaminate all grains and seeds. One of the main functions of a good warehouseman is to so maintain his stocks of grain that none of them are sufficiently invaded by storage fungi to be reduced in grade, price, or processing quality.

When storage fungi invade seeds. Storage fungi do not invade to any serious extent before harvest any of the grains so far tested — wheat, barley, oats, rice, corn, sorghum, soybeans, peas, and common beans. Tuite and Christensen (27) collected ripe heads from different varieties and classes of wheat in several states over a period of three harvest seasons. Some of the plants had been left standing for as long as a month after normal harvest time, or were in shocks or windrows, some of them exposed to frequent rains. From 50 to 100 kernels in each sample were surface disinfected and cultured on malt-salt agar to detect storage fungi. Less than 5% of the kernels yielded storage fungi. Even when NOT surface disinfected or subjected to any other treatment whatever, but merely placed on malt-salt agar, only 4.8% of some 3000 kernels of wheat collected from heads in fields and from combines yielded *Aspergillus glaucus*, the most common storage fungus (see Table 3). Inoculation of ripe plants with spores of storage fungi did not increase the percentage of seeds infected.

Tuite (24) cultured 73,200 surface-disinfected kernels, from 732 samples of soft red winter wheat in Indiana, some from fields that remained unharvested for several weeks after ripening because of heavy rains, and obtained storage fungi from only 25 kernels, or 1 in 300. Tuite and Christensen (26) reported no significant numbers or amounts of *Aspergillus* or *Penicillium* were found in barley kernels before harvest season in the locality where these plants were grown, although the weather was unusually moist, with frequent showers; relative humidities above 75% often prevailed for several days at a time. The grain sampled two weeks after maturity was partly lodged, and thus exposed to conditions that presumably would favor infection of the seeds by fungi. Actually, except for one sample that yielded a number of colonies of *A. flavus* that almost certainly came from contamination during the culturing process, they obtained *no* storage fungi from these surface-disinfected kernels of barley. In tests with undamaged corn picked in 306 fields in Indiana in 1956, 1957, and 1958 Tuite (25) found *Aspergillus* to be rare, and stated, "Aspergilli are uncommon to corn seed in Indiana, prior to harvest." Similarly, Qasem and Christensen (21) collected samples of corn from 20 fields in Minnesota during the rainy post-harvest season of 1957, and found that very few surface-disinfected kernels yielded storage fungi.

In addition to the evidence cited and quoted, we have, over the past fifteen years or so, cultured 50–100 surface-disinfected kernels from each of many hundreds of samples of newly harvested wheat, barley, corn, and beans from many places in the United States, Canada, Mexico, Colombia, and several European countries. Not a single sample has yielded storage fungi from as many as 5% of the surface-disinfected kernels, and the average of freshly harvested, surface-disinfected kernels yielding storage fungi probably would be less than 0.5%, or 1 in 200.

The point is belabored somewhat because it is important. Many men in charge of stored grains have at various times said that sick wheat and related damage was due to something that happened to the grain before it was harvested, and that therefore the warehouseman could not be held responsible for the damage of this sort that developed in his bins. This is absolutely not so. Sick wheat, mustiness, heating, caking, and binburning, all are products of storage fungi and are products of poor storage.

Grain Storage

The data in Tables 3 and 4 indicate that wheat may be contaminated (Table 3, from tests in which seeds were *not* surface disinfected) or invaded (Table 4) to some extent by storage fungi after harvesting but before arrival at the first terminal warehouse.

All the relatively extensive evidence now available, therefore, indicates that wheat at harvest has only a relatively light infection by storage fungi in only a small percentage of the kernels. The great majority of kernels are free of external contamination by storage fungi. This at first seems strange and implausible, but actually it is not; in rather extensive surveys of airborne spores neither we nor others ever have caught many spores of *Aspergillus glaucus*, or *any* spores of *A. restric-*

Table 3. Storage Fungi from Wheat Collected at Various Places from Field to Terminal, Cultured *without Surface Disinfection* on Malt-Salt Agar

			Percentage of Kernels Yielding Fungi	
Sources of Samples	No. of Samples	No. of Kernels	A. Glaucus	Other Storage Fungi
Heads of standing, shocked and windrowed wheat	27	2050	4.8	2.7
Combines	7	1000	4.8	6.8
Country elevators, new crop	16	800	50.0	8.0
Trucks from country elevators unloading at terminal	12	575	64.0	6.0

Source: Tuite and Christensen (27).

Table 4. Percentage of Surface-Disinfected Kernels of Wheat Yielding *Alternaria* and Storage Fungi in Crop Years 1953–54, 1954–55, and 1955–56 as the Grain Arrived at Terminal Elevators

		Storage Fungi			
Region of United States	*Alternaria*	A. *Glaucus*	A. *Flavus*	A. *Candidus*	*Penicillium*
East	71.2	6.6	0.8	0.3	0.8
Southeast	66.6	8.0	3.7	0.5	0.4
Northwest	60.5	10.9	1.1	0.3	0.3
Central	83.4	10.0	1.6	0.1	0.3
South	58.2	9.2	1.8	0.4	0.3
Southwest	67.8	7.5	0.7	0.2	0.1
Pacific Northwest *	48.4	1.4	0.0	0.0	0.0

Source: Kaufmann (17).
  * One year only.

22

tus from air out of doors. On the other hand spores of these fungi may be extremely numerous in the air of some homes, where the fungi themselves are growing in furniture stuffing, pillows, mattresses, paint on the damp basement walls, fabrics, food products, clothing, and leather — materials, that is, whose moisture contents are constantly or intermittently in equilibrium with a relative humidity of 70%–80%. An environment, in other words, similar to that in which many products, including grains, are stored.

As the grains proceed from the farm to terminal, the number of kernels with relatively deep invasion by storage fungi increases, and unless the temperature or the moisture content, or both, of the stored grain is lowered enough to stop the growth of storage fungi, they will continue to grow, and eventually cause germ damage, mustiness, heating, caking, and binburned grain. It is as simple as that.

Source of inoculum of storage fungi. We often are asked why, if grain is relatively or almost free of storage fungi when it is harvested, it is so quickly invaded if stored under conditions that allow the storage fungi to grow. No contradiction is involved. The small amount of inoculum of storage fungi present as dormant spores on the outside of the kernels or as dormant mycelium under the pericarp of a few kernels of even the cleanest lot can, given favorable conditions, increase tremendously within a few days. This overwhelming power of propagation is one of the characteristics of fungi; it is one of the things that enable these fungi to compete so successfully and to take over rapidly in an environment favorable to them. Cases are known in which corn harvested, shelled, and loaded on a truck in the afternoon was by the next morning heating strongly from the growth of fungi.

Mycelium is common and sometimes abundant under the pericarps of most lots of wheat, and it has been suggested that this is mycelium of storage fungi, and provides the inoculum for later rapid increase of these fungi. This may or may not be so. Such mycelium is present under the pericarps of ripe kernels before harvest, kernels from which no storage fungi can be isolated, with or without surface disinfection. Christensen (11) scrubbed wheat kernels (from samples obtained from warehouses, where they had been stored for several months) to remove inoculum from the outer surface, removed strips of pericarp, and cultured them so that they could be observed microscopically, and the colonies which grew out could be later transferred and identified. He stated:

Fungi known to cause deterioration of moist stored wheat were found as dormant spores on the surface of the seed, and as dormant mycelium within the pericarp. The inoculum occurring as external, dormant spores was much more abundant than that in the form of living, internal mycelium. Both external and internal inoculum were more abundant in the low-grade lots than in those of high quality.

Mycelium was present beneath the pericarp of all the seeds examined. Most of this mycelium was dead. In the high-grade lots, most of the living mycelium beneath the pericarp was that of *Alternaria*, a fungus not known to cause deterioration of stored seeds. In the low-grade lots, most of the living mycelium beneath the pericarp was that of *Aspergillus* and *Penicillium* known to be involved in the deterioration of stored grains. No living mycelium was found in seeds more than eight years old.

Grain storage elevators themselves constitute a major source of inoculum of storage fungi, a situation which some warehousemen view with a mixture of exasperation and annoyed disbelief. Culture dishes containing an agar medium favorable to the growth of storage fungi were exposed to outside air in the Red River Valley, at the time of wheat harvest (27). Very few colonies of storage fungi developed on these plates, even after exposure for half an hour, a time sufficient to sample the spore load of more than a cubic foot of air. Culture dishes were similarly exposed to the air within country elevators in the same region and on these dishes, exposed for 30 seconds to three minutes, hundreds to thousands of colonies of storage fungi developed, including especially *Aspergillus restrictus*, *A. repens*, *A. ruber*, and *A. candidus*. These colonies grew from spores produced by storage fungi growing within the elevator itself, most probably, of course, from fungi growing on moist stored grain in the elevator.

We also have cultured numerous samples of dust collected from grain elevators and terminal warehouses, and have consistently found a gram of such dust to yield from several hundred thousand to several million colonies of storage fungi. So there is no mystery about inoculum — there is plenty of inoculum of storage fungi on and within the grains at harvest, and a superabundance of it in the warehouses themselves. This means that practical control of these storage fungi lies not in the wholly impossible task of reducing or eliminating inoculum, but rather in maintaining the grain during storage under conditions that do not permit the storage fungi to develop sufficiently to cause damage. This is the topic of the next section.

CONDITIONS THAT FAVOR DEVELOPMENT
OF STORAGE FUNGI

The major conditions that influence the development of storage fungi on stored grains are as follows: (1) the moisture content of the stored grain; (2) the temperature; (3) the length of time the grain is stored; (4) the degree to which the grain already has been invaded by storage fungi before it arrives at a given site; (5) the amount of foreign material present in the grain; and (6) the activities of insects and mites.

Each of these is closely related to most of the others. Many lots of wheat and corn, for example, are stored at moisture contents high enough for storage fungi to grow in them; a safe moisture content for storage of a given lot will depend in part on the temperature of that lot of grain in store, the length of time it is to be held, and whether it is almost free of storage fungi or already partly invaded. Thus, while each of these factors will be discussed separately, the reader must recognize that they are related, interacting, and inseparable. To the extent that data on these factors are accumulated on a given stock of grain and applied to decisions concerning handling and storage, science replaces art and knowledge replaces guesswork in grain storage.

Moisture content. For each of the common species of storage fungi there is a minimum moisture content in grain below which it cannot grow. These minimum moisture contents have been determined fairly accurately for most of the common storage fungi growing on the starchy cereal seeds and on some of the oil seeds, and are summarized in Table 5.

The most drouth-resistant of the storage fungi, *Aspergillus restrictus* and *A. halophilicus*, cannot grow at moisture contents below those in equilibrium with a relative humidity of approximately 65%. Any seeds whose moisture content is below that in equilibrium with a relative humidity of 65% (see Table 5) should therefore be safe from invasion by storage fungi, regardless of the other conditions of storage.

If wheat is stored at a moisture content of 14.0%–14.5% and at a temperature of about 70° F. it will be slowly invaded by *A. restrictus*, the rate of invasion and the damage done being proportional to the moisture content (much more rapid invasion, much more damage, after a year's storage at 14.8% moisture content than at 14.0%); to the temperature (much more rapid invasion, much more damage in wheat of 14.5% moisture content stored at 90° F. than at 45° F.); and to the time of storage (even in the wheat free of storage fungi with 14.8% moisture content there will be no damage sufficient to cause grade reduction if stored at 70° F. for six months, but there will be after a year). And it

Table 5. Moisture Content, Wet Weight Basis, of Several Common Grains and Seeds in Equilibrium with Relative Humidities of 65%–85% at 20°–25° C. (in Percentage)*

Relative Humidity	Wheat and Corn	Rice		Soybeans	Sunflower Seeds†
		Rough	Polished		
65%	12.5–13.5	12.5	14.0	12.5	8.0
70%	13.5–14.0	13.5	15.0	13.0	9.0
75%	14.5–15.0	14.0	15.5	14.0	10.0
80%	16.0–16.5	15.0	16.5	16.0	11.0
85%	18.0–18.5	16.5	17.5	18.0	13.0

* The figures given can be only approximate, since the equilibrium moisture contents of a given kind of seed vary with several factors such as variety and location, and especially with whether the grains or seeds are absorbing or losing moisture to attain the equilibrium.

† Oil type; the confectionary and birdseed types have equilibrium moisture contents about 1.0% above these.

is *only* at moisture contents below 15.0% that *A. restrictus* will predominate in wheat stored for some months (12, 28). At moisture contents (in wheat) above about 15.0% other species of the *A. glaucus* group, such as *A. repens, A. amstelodami,* and *A. ruber,* predominate and may maintain their predominance even at moisture contents up to 18%.

Not uncommonly we receive samples of wheat from a commercial bin in which enough germ damage has developed to cause severe reduction in grade. When surface disinfected and cultured on an agar medium containing 7%–10% NaCl, 90% of the kernels yield *A. restrictus.* This is conclusive evidence that the grain was stored for at least some months at a moisture content of approximately 14.5%–14.8%; otherwise *A. restrictus* just could not have developed to that extent. The warehouse records on this lot of grain may show that it had a moisture content of 12.2%; this means only that the average sample taken from the lot of grain while it was being binned had a moisture content of

12.2%, as determined by the combination of machine, man, and method used to determine moisture content.

A. repens, A. amstelodami, and *A. ruber,* all members of the *A. glaucus* group and so closely related that only the expert can or needs to distinguish one from another, invade wheat and corn stored at moisture contents of 14.5% or 15.0% to 17.0%. If inoculated onto a sample of wheat from which all competing storage fungi have been eliminated, some strains of these species can grow at moisture contents almost as low as can *A. restrictus.* But if a mixed population is present, as is inevitable on wheat in bins, members of the *A. glaucus* group seldom can predominate at a moisture content below about 14.5% or 15.0%, and at moisture contents above 17% they are replaced by still other species. So if we culture surface-disinfected kernels from a lot of wheat from a warehouse, and these species of *Aspergillus* grow from most of them, it is biologically irrefutable evidence that the grain had been for some time at a moisture content of at least 15.0% to 15.5%, regardless of what the warehouse records show.

These three common species also form small yellow spherical fruit bodies, technically known as perithecia, on the surface of the seeds they have invaded, or on the surface of the embryos. In the starchy cereal seeds, including wheat and corn, these perithecia are formed only if the moisture content of the grain is in excess of about 15.5%, and in practice they are likely to be formed on grain whose moisture content is between 15.5% and 16.5% only. If the moisture content is higher than 16.5%, fungi other than these are likely to take over. Thus if we encounter a lot of wheat and corn, as we occasionally do, with many perithecia of these fungi on the surface of the embryo (those on the outside of the seed were scoured off as the grain was loaded out of the bin) we can be positive that the grain was stored for at least six weeks at a moisture content of 15.5% or above, most likely between 15.5% and 16.5%, and at a temperature of 65° F. or above — or for a considerably longer time at that moisture content but at a lower temperature. It is *only* under such conditions that these fungi will form perithecia.

Similarly, if we culture surface-disinfected kernels of wheat from a warehouse and *A. candidus* grows from a considerable number of them, we know that the seeds *had* to have been stored at a moisture content above 15%, and most likely at a moisture content of 16%–17%, because otherwise *A. candidus* could not have invaded them. And if the seeds are invaded by *A. flavus* they had to have had a moisture content of at least 18%, more likely close to or above 19%.

We have received samples from bins in which a high percentage of germ damage or sick wheat had developed and which the warehousemen protested must have been of mysterious origin because none of the grain ever had a moisture content in excess of 12.37%. When surface disinfected and cultured, these samples yielded *A. candidus* and *A. flavus* from a large percentage of the kernels, proving that the moisture content as given on the warehouse records was in error by some 3% to 6% or 7%.

In the range of moisture content between 14.0% and 15.5% in the starchy cereal seeds, or between 13.0% and 14.0% in soybeans, or between 10.5% and 11.5% in flaxseed, a difference of only 0.2% in moisture content makes a great difference in the rate of growth of storage fungi and in the damage they do in a given length of time. This is a smaller difference than can be measured accurately by moisture meters. If a large bulk of wheat or corn were to be stored for months with an average moisture content of 14.0%, it would seem only a common-sense precaution to remove samples from different portions at intervals and test them for moisture content, germ damage, and storage fungi; in this way the warehouseman knows at all times what is going on in all parts of the bulk.

Temperature. The storage fungi common on grains grow most rapidly at about 30°–32° C. (85°–90° F.), and their growth rate decreases as temperature decreases. Some strains of the *Aspergillus glaucus* group grow slowly at temperatures of 35°–40° F., and some species of *Penicillium*, which require a higher moisture content than the drouth-resistant species of *Aspergillus*, can grow at a temperature several degrees below freezing.

Qasem and Christensen (21), working with samples of corn stored in the laboratory at moisture contents of 16% and 18% and temperatures of 5°, 10°, 15°, 20°, and 25° C., found that deterioration — as measured by invasion of the seed by storage fungi, decrease in germination percentage, and increase in percentage of discolored germs — was proportional to moisture content, temperature, time, and degree of invasion by storage fungi at the beginning of the tests. They state,

"Low temperature was as effective as low moisture content in preventing damage by the fungi tested."

Papavizas and Christensen (20), studying the fungus invasion and deterioration of wheat stored at low temperatures and moisture contents of 15%–18%, found that "wheat with a moisture content up to 16% may be stored without obvious deterioration for a year at a temperature of 10 C or below, and wheat with a moisture content up to 18% may be stored safely for as long as 19 months at a temperature of —5 C." They worked with wheat that was relatively free of storage fungi at the beginning of the tests; it would be desirable to have data from similar tests with wheat that at the beginning had been invaded to various degrees by storage fungi.

The preserving effects of low temperature have, of course, long been known, and the principle of refrigeration has long been applied to preservation of many perishable products. As indicated in Chapter 8, preservation of quality in stored grains by maintenance of low and uniform temperature is a relatively recent development.

The effect of temperature on the growth of storage fungi and on the damage they do has some complexities too. Corn or wheat that has not been invaded by storage fungi to any serious degree or extent, and is otherwise sound and in good condition, can be stored at a moisture content of 15.0% for nine months to a year without damage even if the temperature is 45°–50°F. If, on the other hand, the grain already has been moderately to heavily invaded by storage fungi, and is stored at a moisture content of 15.0% and a temperature of 45°–50° F., the fungi may continue to grow, and within six months may cause extensive spoilage. This is another case where information on the number and kinds of fungi on grain going into long-time storage greatly aids a warehouseman in evaluating grain condition and storability.

Length of time the grain is stored. If grain is to be kept for only a few days, its condition and storability may be of minor interest, at least to the seller, although they may be of more interest and importance to the buyer. The grain standards now in force in the United States were evolved to give a reasonable assurance that a lot, say, of wheat or corn conforming to the specifications for grade No. 2 would remain in good condition for a reasonable length of time, although certainly few people familiar with long-time storage of corn would consider that the 15.5% moisture content maximum specified for grade No. 2

would permit storage without fairly rapid spoilage. What constitutes a "reasonable length of time" is a good question.

Some lots of grain are stored for years. When the moisture content limits for different grades of various kinds of grain and seeds were first established it evidently was assumed that what was good for a few weeks or a few months would be good forever. So far as can be determined now, the moisture content limits for some of the grains were established at least partly on the basis of respiration tests (actually, CO_2 production) of relatively short duration. However, no one has shown that *Aspergillus restrictus*, growing in wheat or corn with a moisture content of 14.5%, will respire measurably even over a period of several months.

We now know that a moisture content that is safe for storage of a given lot of grain for two weeks may not be safe for two months, and a moisture content that may be safe for storage of a given lot for two months may not be safe for two years. It sometimes has been thought that the longer the storage period is to be, the lower the moisture content should be. This is true within certain limits and with some reservations. A much better rule would be to know accurately the average, *and the range*, in moisture content of the grain in each bin or car or shiphold, and also the numbers and kinds of fungi — when the grain is first loaded in, and periodically thereafter. This would enable the warehouseman to know at all times the condition and the deterioration hazard of grain throughout his warehouse.

One often hears statements to the effect that "we store flax at 11% moisture content without difficulty," or "in our region farmers and elevator men store wheat with a moisture content of 15% without any trouble." Of course they do, but the catch is that difficulty or trouble may simply not be visible but is there and is building up for whoever next stores it. As described in Chapter 1, flaxseed with a moisture content of 11.0% may or may not be safe for storage. The same is true of other kinds of seeds. We can hold wheat at 15% moisture content in the laboratory for two months at a temperature of 70° F. without germ damage developing — if the wheat is originally almost free of storage fungi and otherwise sound. After two months at 15% moisture content and 60°–70° F., however, the wheat is moderately to heavily invaded by storage fungi, it is an exceedingly poor storage risk, and unless dried to a moisture content of 13% or below it is almost certain to develop a rather high percentage of germ damage. It looks sound, but it is not, and whoever bought it would be buying trouble.

The degree to which grain already has been invaded by storage fungi. If grain immediately subsequent to harvest has been so stored that it has been invaded by storage fungi, it is already in the first stages of deterioration. Such grain, if later stored under conditions that permit spoilage to continue, will develop more damage in a given amount of time, or will develop a given amount of damage in a shorter time, and will be subject to continued invasion by storage fungi and associated damage at a lower moisture content and lower temperature, than perfectly sound grain. This is normal in microbiological spoilage. Fabrics that are partly weakened by fungus invasion will, if exposed to conditions that permit this decay to continue, be rotted to the point of failure more rapidly than perfectly sound fabrics. Wood partly decayed by fungi — and fungi are the *only* cause of decay in wood — will, if exposed to conditions that promote decay, fail proportionately faster than perfectly sound wood. Spoilage of stored grains by fungi does not differ in principle from decay of wood: the fungi are different, the conditions different, especially in that the species of *Aspergillus* mainly responsible for germ damage or germ decay in grains can grow at a much lower moisture content than can the fungi that cause wood decay (see Plate 5); otherwise the processes are basically alike.

Qasem and Christensen (22) found that corn already invaded to some degree by storage fungi deteriorated much more rapidly when stored under conditions that permitted the fungi to grow than did sound corn that at the start of the tests was free of storage fungi.

Sauer and Christensen (unpublished data) collected samples of wheat at terminals, stored them at the same moisture contents as they had when received and at 75° F., and periodically tested them for moisture content, germination percentage, discolored germs, and number and kinds of fungi. Some of the samples when received yielded storage fungi from about 30% of the surface-disinfected kernels and *Alternaria* or other field fungi from 50%. These samples decreased in germination percentage more rapidly than samples at the same moisture content (13.9%–14.1%) which originally yielded storage fungi from none of the surface-disinfected kernels and field fungi from over 90%. Also the samples originally moderately invaded by storage fungi developed in 75 weeks up to 12% dark brown germs with an additional 8% light to medium brown germs, enough to put them into sample grade, while the samples which originally yielded storage fungi from none of the surface-disinfected kernels developed no dark brown germs and still retained the characteristics of grade No. 2 hard red winter wheat.

If grain at harvest is almost free of storage fungi, one would expect all lots of grain, at a given moisture content, to start off more or less even after harvest. They do. However, very soon different lots, with different moisture contents and different storage histories are mixed together, and sometimes grain from a previous year's crop is mixed in with a parcel of this year's grain. Grain that has been stored for a year may be just as sound as freshly harvested grain; on the other hand it may be partly deteriorated, which is one obvious reason for the seller wanting to mix it in with a good lot and get rid of it. This deterioration may not have progressed to the stage of germ damage, but involves rather an invasion of a considerable percentage of kernels by storage fungi. Such grain is less desirable for long-time storage than is sound, freshly harvested grain free of storage fungi, and it can be detected readily by simple laboratory tests. We once tested freshly harvested truckload lots of corn as they came in from fields to a country elevator, and found them essentially free of storage fungi. The next day samples were taken from truckload lots being loaded out of this elevator and bound for a terminal; storage fungi grew from approximately 50% of the surface-disinfected kernels. The only possible explanation is that some corn partly invaded by storage fungi, partly deteriorated but still grade No. 2 yellow corn, was being mixed in with the freshly harvested corn. The aim may well have been to mix dry corn with moist corn to achieve an average moisture content specified for grade No. 2, but in practice corn of low moisture content, heavily invaded by storage fungi and already partly deteriorated, was mixed with sound corn of high moisture content. This should assure a high storage risk and a short storage life for the mixed lot unless it was soon dried well below the upper limit of 15.5% specified for grade No. 2 corn.

In grade No. 2 yellow corn, there may be a very wide range in storability. We have sampled some bins of 100,000 bushels each of No. 2 yellow corn in which none of the samples taken from the bin had a moisture content over 14.2%, there was no germ damage, none of the kernels were invaded by storage fungi, and even germination (not considered in grade) exceeded 95%. This corn was essentially of seed grade and, by the way, it was kept in storage for four years without any damage or any decrease in grade.

We have sampled other lots of No. 2 yellow corn in which the moisture content was 15.2%, slightly below the allowable maximum, 80%–90% of the surface-disinfected kernels yielded storage fungi, germ damage was 5% (the maximum allowable in grade No. 2), another 20%–30% of the kernels had slightly discolored (and dead) germs which, at any temperature over 60° F. and at a moisture content as high as this corn had, would soon become brown, and the germination was 40%. Kept at 25° C. (77° F.) for several months, this lot developed very heavy damage from fungi, especially "blue-eye" from masses of spores of *Aspergillus glaucus* produced on the surface of the germ under the pericarp. Samples stored at 20° C. (68° F.) developed somewhat less, but still extensive, damage.

That is, one of the lots described above was sound, the other was partly deteriorated. Both were No. 2 yellow corn and commanded the same price. One could be stored without risk of deterioration, the other could not be. Simple laboratory tests would serve to distinguish such lots easily.

Foreign material. Foreign material consists mostly of particles finer than the seeds themselves, broken seeds, weed seeds, fragments of plants, parts of field insects such as crickets and grasshoppers, and soil. When grain is loaded into a bin from an overhead spout, this fine material usually accumulates below the spout, and at times may almost fill the interseed spaces in a column of grain there. This material is an excellent breeding ground for fungi and for some kinds of insects and mites. If it is tightly packed, air will flow around it when a bin is aerated to reduce the temperature, and spoilage may begin in such places. For long-time storage, the less foreign material in the grain the better.

Insects and mites. Insects and mites affect the development of storage fungi by (1) increasing moisture content of the grain and (2) carrying spores of fungi into the grain. Like all living things, insects and mites break down much of their food into carbon dioxide and water, and so they are likely to increase the moisture content of the grain in which they are active. The storage fungi do so also, but at a much slower rate. In our laboratory tests the increases in moisture content as a result of storage fungi growing for a year or more in grain originally at a moisture content of 15%–16% have been on the order of 0.2%, barely large enough to be detectable. This may not be true of the more rapidly growing species such as *Aspergillus candidus* and *A. flavus*, which also require a higher moisture content to begin to grow, but it certainly is true of *A. restrictus* and *A. repens.* Agrawal *et al.* (8) stored grain of 12.1% moisture content at 75% relative humidity, infested and not infested with granary weevils, *Sitophilus granarius.* After three months the grain *not* infested with weevils had a moisture content of 14.6%–14.8%, the amount expected in wheat in equilibrium with a relative humidity of 75%, while the moisture content of the samples infested with weevils ranged from 17.6% to 23.0%. They stated that "— rather forceful aeration was required (with air at a relative humidity of 75%) to maintain a uniform moisture content when the insects were increasing." Others also report rather large and rapid increase of moisture content in grain infested with insects (13, 19). Sinha (23), in a study of hot spots in farm-stored grain in Manitoba and Saskatchewan, Canada, found grain in the center of a hot spot, where the insects were most active and the temperature highest, to have a moisture content of 14.6%–17.0% (the grain not infested had a moisture content below 14.5%) while the grain above the hot spot had a moisture content of 17%, and grain in the surface layer was germinating, which means a moisture content of at least 25%. He found as many as 12 different species of insects and mites in a single hot spot of wheat, and 16 species in one hot spot of oats.

A case is reported (9) in which an infestation of mites raised the moisture content of flour from 15.4% to 28.10% in eighteen months. Microbes probably were involved, although no tests were made to determine this. No data are available concerning increases in moisture content of grains infested with mites, but mites sometimes are very numerous in stored grain and their activities almost certainly would be accompanied by some increase in moisture content.

Agrawal *et al.* (8) report a "close and constant" association between granary weevils and storage fungi, especially *Aspergillus restrictus*, and several others report that increases in storage fungi regularly accompany infestations of grain by insects (13, 19). There seems to be no doubt that at least some of the common grain-infesting insects regularly carry into the grain they infest a large load of inoculum of storage fungi and, as they develop in the grain, provide conditions favorable for the development of these fungi. Therefore what appears to be an insect problem may be an insect-plus-storage-mold problem. Fumigation may kill the insects, but have little or no effect on the storage fungi (13), and so the damage from storage fungi may continue.

Less work has been done on the relation between grain-infesting mites and storage fungi than that between insects and storage fungi, but such evidence as has been accumulated from laboratory studies and observation indicates that mites and storage fungi go together. All the mites that we have observed in samples of stored grain require a moisture content as high as that which permits some of the storage fungi to grow. Griffiths *et al.* (16) stated: "Grain-infesting mites *Acarus siro* and *Tyrophagus castellanii* were found in some abundance in samples of commercially stored wheat, the moisture content of which ranged from 13.5 to 15.0%, which is the range of moisture content at which fungi in the *Aspergillus glaucus* group are likely to predominate. These mites, when developing in moldy grain, picked up spores of the storage fungi and carried these spores on the outside of their bodies, in their digestive tract, and in their feces. As they entered clean grain they inoculated it heavily with spores of these fungi and later they fed to a considerable extent upon the fungi that developed." Also, "it is likely that whenever heavy infestations of these mites are found in grain, there will be damage not only from the mites feeding on the embryos of the kernels, but also from the storage fungi that accompany them."

3 MEASUREMENT OF MOISTURE CONTENT

Accurate knowledge of the amount of water in grains and seeds is important for two reasons: (1) in merchandising, because the water is being bought and sold at the same price as the grain of which it is a part; (2) in storage, because moisture content is a major factor in storability, chiefly because of its influence upon the growth of storage fungi. The chapter "Moisture and Its Measurement" in the book *Storage of Cereal Grains and Their Products* (29), published in 1954, consists of 42 pages of text and 93 references, which indicates the importance that cereal chemists attach to accurate measurement of moisture in grains and grain products. When that chapter was written in the early 1950's, some of the major sources of error in measurement of moisture in different portions of large bulks of grains were not known, and virtually nothing was known about the rate and magnitude of shifts of moisture, with time and with differences in temperature, from place to place in a bulk of grain.

At that time, practical grain men assumed that they knew the moisture content of the grain in their warehouses, and others, including those who were investigating deterioration of stored grains, also assumed that they did. The moisture content of a representative sample, taken as the bin was being filled, was thought to be the moisture content of *all* the grain in the bin, and the moisture content was thought to remain uniform throughout the bulk throughout the storage life of the grain. It was only after samples had been taken from grain in different portions of a number of bins at different times and at different places, and tested for moisture content, that the fallacy of this idea was exposed. Competent warehousemen now recognize that in a bulk of stored grain moisture may shift from place to place, sometimes rapidly, but in the early 1950's this was not known. Even now many warehousemen still believe that they can know the moisture content of the grain under their care by looking at the figures in the ware-house records — which give the average moisture content of represen-tative samples — rather than by withdrawing samples from different portions of the bulk and testing the moisture content of each sample separately. The present chapter is devoted chiefly to a discussion of the difficulties in knowing accurately the moisture content of any given portion of a given bulk of grain at any given time, and to means of overcoming these difficulties. Several case histories are given in some detail, to illustrate the consequences that ensued when moisture contents were not what they were assumed to be.

MOISTURE CONTENT AND GRAIN STANDARDS

Moisture content is one of the characteristics specified for the dif-ferent grades of different kinds of grain. Naturally, if the grade and price of a given lot of grain are to be based on a sample taken from that lot, the sample must be representative of the lot from which it is taken, and a good deal of work and thought has gone into methods of sampling to ensure that the samples will be truly representative of the lots from which they are taken.

The specification of rather precise methods of sampling, and of pro-cedures for measuring various characteristics, including moisture con-tent, of the sample taken, and the establishment of rather precise lim-its of moisture content for different grades, have led warehousemen to assume (1) that the moisture content of the lot from which the sam-ple was taken is uniform; (2) that the moisture content of the sample can be accurately measured; (3) that the moisture content during storage will remain at least fairly constant; (4) that the upper limit of moisture content specified for the highest grade of a given kind of grain, or for the grade most commonly encountered in commerce, such as grade No. 2 corn, is low enough so that the grain of that moisture con-tent will be safe from deterioration for a reasonable length of time in storage. We now know that in some lots of grain none of these assump-tions is correct. Grain standards, as mentioned in Chapter 1, were established to promote orderly marketing. Storability is another matter.

MOISTURE CONTENT AND GRAIN STORAGE

For maintenance of quality in storage the average is meaningless. Rather it is essential to know *whether any considerable portion of the lot or bulk in question has a moisture content high enough to permit damaging invasion by storage fungi under the conditions likely to be encountered during the storage life of that bulk.* The figure on average moisture content of the lot not only will not tell us this, but will at times give an entirely false picture of the storability of the grain. One of the responsibilities of an elevator superintendent or warehouse-man is to recognize lots of grain of high storage risk, and to handle them so as to minimize losses during storage. To evaluate storage risk, he must know the *range* in moisture content in different places of each of the bulks under his charge. It is of small comfort and little use to him to know that the average moisture content in a given bin of wheat is 13.0% if some of the grain has a moisture content of 16.0% and is heating and musty, and other portions of the grain have a moisture content of 10.0%. Before forced aeration systems were developed for maintenance of uniform temperatures and moisture contents in large masses of grain, spoilage due to exactly such unequal distribution of moisture through the grain mass was common, although the reason for it was not known. It still is common in some unaerated bins.

To many warehousemen, grain merchants, and grain processors, the moisture content of a representative sample of grain *is* the moisture content of the entire lot or bulk. They envision the moisture as uniformly distributed through the mass, with every kernel having about the same moisture content as every other kernel. And regardless of how long or under what conditions the grain is stored, the moisture content of the representative sample, as recorded when the grain was loaded into the bin, is, in their minds, the moisture content of the grain throughout its storage life.

Because of these concepts — which are grossly erroneous — many of those in charge of stored grains, from the farmer with a few hundred to a few thousand bushels to the superintendent of a terminal warehouse in charge of hundreds of thousands or millions of bushels, have believed and maintained that, under some conditions or circumstances, wheat or corn stored with a moisture content of 12%–13% would mysteriously "go out of condition." That is, it would develop extensive germ damage, and would heat and become musty, with consequent loss of grade or even complete loss of the grain and sometimes of the warehouse as well.

Some still believe that corn or wheat of 12%–13% moisture content sometimes will heat and spoil. What they really mean, although it is seldom so expressed, is that the grain in their bins may spoil when, according to the warehouse records, it has a moisture content of only 12%–13%. It may or may not have had an *average* moisture content of 12%–13% when loaded into the bin, but the portions of the grain where spoilage occurred certainly had a moisture content considerably above that for some time before spoilage began. We have taken samples from different places in many bins and determined the moisture content of each of these separately. We commonly have found differences as large as 2%–3%, and sometimes as large as 5%–10%, between moisture content of this grain as shown on the warehouse records and the actual moisture content of the individual samples. The major reasons for discrepancies such as this will now be discussed.

ERRORS IN MEASUREMENT OF MOISTURE CONTENT

The sample. To determine moisture content of grains by means of the electric moisture meters in common use, a sample of 250 grams is taken. For the purposes of illustration this may be said to equal 0.5 pound. Various methods of removing grain from trucks, railway cars, bins, etc., are used to obtain representative samples, but the details of these methods and procedures need not concern us here. The point is that the final sample on which the moisture content actually is determined amounts to only half a pound. If this sample is taken from a bulk of 1000 bushels of corn weighing 54 pounds per bushel, the sample constitutes only 108,000th of the bulk from which it was taken. If it is taken from a mass of 10,000 bushels, the sample amounts to only 1/1,080,000th of the total bulk from which it was taken. Such a sample may be *representative* of the bulk, in showing the *average* moisture content, but it cannot conceivably indicate the range in moisture content of the mass from which it was taken. Even in grain as uniform as can be encountered in practice, we find a range of 0.5% in moisture content among several samples taken from a truckload lot, and we commonly find a range of 1.0 in moisture content among samples taken from a carload. For maintenance of quality in storage, insistence upon a carefully worked-out system of sampling, followed by a precise determination of moisture content to the second place to the right of the decimal point, puts the emphasis entirely in the wrong place. Several samples taken from different portions of the mass, with each tested separately for moisture content, will give much more information on storability than will standard sampling procedures. The two procedures or approaches are not mutually exclusive, of course. Both can be used, the moisture content of the representative sample in marketing and the moisture contents of a number of individual samples to evaluate storability.

In bins of grain that have been stored for a few weeks to a few months, a range in moisture content of 2%–3% and sometimes more is not at all unusual when several samples are taken from different portions and tested separately. One example will be cited to illustrate this. In 1952 we obtained samples from depths of three, six, and nine feet in a bin of hard winter wheat in Chicago, several months after the bin had been filled. The bin was a quonset type, and held 300,000 bushels. All the grain that went into this bin had been mixed thoroughly twice before being put into the bin, and the moisture contents of representative samples of all lots had been measured carefully. The superintendent in charge of this terminal was an experienced and able warehouseman; among other things he even checked the accuracy of his moisture meter daily by oven drying several samples, certainly a most unusual procedure at that time. He assured us that no grain in this bin had a moisture content above 14.0%. We determined by oven drying the moisture content of the eighteen samples we had taken, and stored the remaining portions in closed bottles to see what would happen. Approximately half the eighteen samples had moisture contents above 14.0%, and several of them had moisture contents above 15.0%. After they had been kept in the laboratory for several months, storage fungi grew on all those with moisture contents above 14.0%, and eventually considerable germ damage developed.

Methods of measuring moisture content. Precise measurement of moisture in grains, as in many other materials, is difficult. This is well recognized by investigators, if not always by practical warehousemen. The Official Grain Standards of the United States (7) specify that "Moisture shall be ascertained by the air-oven method prescribed by the United States Department of Agriculture, as described in Service and Regulatory Announcement No. 147, issued by the Agricultural Marketing Service, or ascertained by any method which gives equivalent results." In practice, moisture content is determined by means of electric moisture meters, and oven drying is used only to check the accuracy of the meters. Even the oven methods are not infallible. As pointed out by Zeleny (34), "all oven methods for determining moisture in grain are more or less empirical in nature, the results depending on the degree of subdivision of the material being tested, the time of drying, and the temperature and atmospheric pressure under which the drying is accomplished. In most biological materials, including grain, it is difficult if not impossible to remove all moisture by the application of heat without at the same time driving off small amounts of other volatile substances or causing decomposition of some of the constituents with the formation and release of moisture not initially present as such."

However, the standard methods for determining moisture contents of common grains, seeds, and grain products by oven drying, if not absolutely exact, are at least relatively accurate. If a given procedure, such as drying a sample of the grain at 103° C. for three days, indicates that wheat with a moisture content of 13.5% is safe from invasion by fungi while wheat with a moisture content of 14.5% is invaded by fungi, we can be fairly sure of close agreement in the results of different workers in different places at different times concerning the limits of moisture content that permit invasion of wheat by fungi.

As stated above, in practice in the grain trade, moisture content is determined by means of electric meters, because speed is essential. Until July 1962, the Tag-Heppenstal or Weston meter was the standard one in grain inspection offices. Although moisture content as determined by this meter commonly was expressed to the second place to the right of the decimal point, on some samples in our tests it gave readings as much as 1.0% to 1.5% low, and so was in error to the first place to the left of the decimal point. Expressed in another way, the error at times was 100 to 150 times as great as the supposed sensitivity of the meter. In 1963–64 the Motomco became the standard meter in the grain inspection offices of the United States Department of Agriculture. In our tests with many samples of different kinds of grain this meter has given readings in very close agreement with those obtained by oven drying, in over 90% of the samples. Other brands of meters now in general use evidently are equally accurate. Typical results of tests of moisture content with four kinds of meters and by oven drying are shown in Tables 6 and 7.

Accurate measurement of moisture content by means of a meter requires accurate measurement of the weight of the sample of the grain, which requires an accurate balance, and also requires accurate measurement of the temperature of the grain, which requires an accurate thermometer, and all three require an accurate operator. It is only common sense to check the accuracy of the meter occasionally, either by determining the moisture content of check samples through oven drying, or by submitting samples to grain inspection

Table 6. Moisture Contents as Determined by Meters and by Oven Drying of Hard Red Winter Wheat, 1960 Crop, in St. Louis (in Percentage)

Sample	Meter				Oven Drying
	Motomco	Radson	Steinlite	Weston	
1	14.20	14.20	13.31	13.51	14.2
2	13.70	13.40	13.85	12.70	14.3
3	14.30	14.00	13.44	12.76	14.5
4	13.30	14.60	16.35	13.97	14.5
5	14.77	14.50	15.00	13.52	15.1

Table 7. Moisture Contents as Determined by Meters and by Oven Drying of Soft Red Winter Wheat, 1961 Crop, in New Orleans (in Percentage)

Sample	Meter				Oven Drying
	Motomco	Radson	Steinlite	Weston	
1	13.04	13.50	13.05	11.75	12.5
2	13.99	14.50	14.20	13.40	13.6
3	14.34	14.70	14.50	13.43	13.8
4	14.20	14.50	14.70	13.43	13.9
5	14.90	14.90	15.05	14.00	15.5

offices or to laboratories for such determination. If the moisture content of a given lot of grain is near or within the range where deterioration might occur, several samples should be taken and their moisture contents determined.

Moisture transfer. In any large mass of stored grain that is not artificially aerated, and in some not-so-large masses, slow movement of air, set up by differences in temperature between different portions of the bulk, results in moisture transfer. This is described in some detail in Chapter 8. Shifts of moisture content are likely to be large and rapid in grain bulks in which there are relatively large differences in temperature and in which the moisture content is high, but they are not limited to such conditions. The results of tests in a storage bin at Richmond, Va., will illustrate this moisture shift.

Christensen and Drescher (30) buried samples of wheat in small cloth bags, each bag containing about two pounds of grain, in a bin of grain as the bin was being filled with wheat. The wheat in the bags had a moisture content of 11.0% and a germinability of over 95%, and it was almost free of invasion by storage fungi. This was seed-quality hard red spring wheat that had been grown in the dryland region of Montana. The bin was filled with wheat that had an average moisture content of 13.2%. Each sample bag was numbered, and its location in the mass recorded. Within three months after the bin had been filled, heating developed in the grain in various places, and so the bin was emptied. The small bags of wheat were recovered from the grain as it came out of the hopper, the contents of each bag were placed in a moisture-proof tin, and the tins were sent to the laboratory. There they were tested for moisture content, number and kinds of fungi, germ damage, and germinability.

The *average* moisture content of the grain in the 36 bags recovered was 13.2%, identical with the average moisture content of the mass in which they had been buried. The *highest* moisture content of the grain in one of the bags was 18.0%, and the lowest moisture content of the grain in another of the bags was 10.0%. The number and kinds of fungi present in the grain in the small bags indicated conclusively that the grain in nearly all the bags had, at one time or another during the three months of storage, moisture contents in excess of 15.0%. Severe germ damage had developed in some of the samples. All this evidence proved beyond question that there had been large and rapid shifts in moisture content in the grain in this bin. Without the evidence gained from these samples, there would have been no rational explanation of why the grain in this bin had begun to heat and spoil while stored with an average moisture content of 13.2% — it would have been just another example of "mysterious" spoilage.

In another case, a large bin of wheat in a terminal elevator began to heat, although it had, according to the warehouse records, a moisture content of 12.2%. While storage fungi can slowly invade wheat that has a moisture content of about 13.5%, the fungi cannot grow rapidly enough to cause heating unless the moisture content of the grain is above 15.0% or 15.5%. It was supposed therefore that something other than storage fungi must be involved in the heating.

The bin was emptied, and samples were obtained from the grain as it flowed out of the hopper at the bottom. These samples were taken to the laboratory and tested immediately for moisture content and for numbers and kinds of fungi. Some of the samples had moisture contents over 16.0%. Since grain flowing out of a bin through a hopper at the bottom undergoes considerable mixing, much of the grain in this bin probably had a moisture content above 18.0%. In fact, the fungi that grew from the surface-disinfected kernels on agar indicated that the moisture content of some of the grain was over 18.0%; otherwise those kinds of fungi could not have invaded it. The moisture content as shown in the warehouse records was not the moisture content of much of the grain in the bin.

Such cases as described above are not rare, although they are less common than they once were, because of the use of forced aeration to maintain a moderately low and uniform temperature and a uniform moisture content through bulks of stored grain. The major point to be got across here is that unless the man in charge of a given lot of stored grain occasionally removes samples from the bulk and tests the individual samples separately for moisture content, he has no idea of the moisture content that actually prevails in different portions of the grain in a given bin. Stated in another way, the moisture content of stored grain can be determined only by examining samples taken from the grain, not by examining the warehouse records. Good storage practice often is as simple as that.

Mixing and blending. Almost inevitably there is some mixing of grain of different moisture contents from harvest on. Wheat and other small grains combined in the early morning may have a moisture content 5% above grain in the same field combined in mid-afternoon. It also is a common practice, from the farm to the final processor of grains, to mix different lots of grains of different moisture contents to achieve an average moisture content that is specified for a given grade or by a given customer or that is thought to be safe for storage.

The history of one such series of mixings or blendings is shown in the diagram on page 45. In this case, the purpose of mixing the lots of different moisture content was to reduce the moisture content of the originally moister lots, and to make the whole safe for storage for at

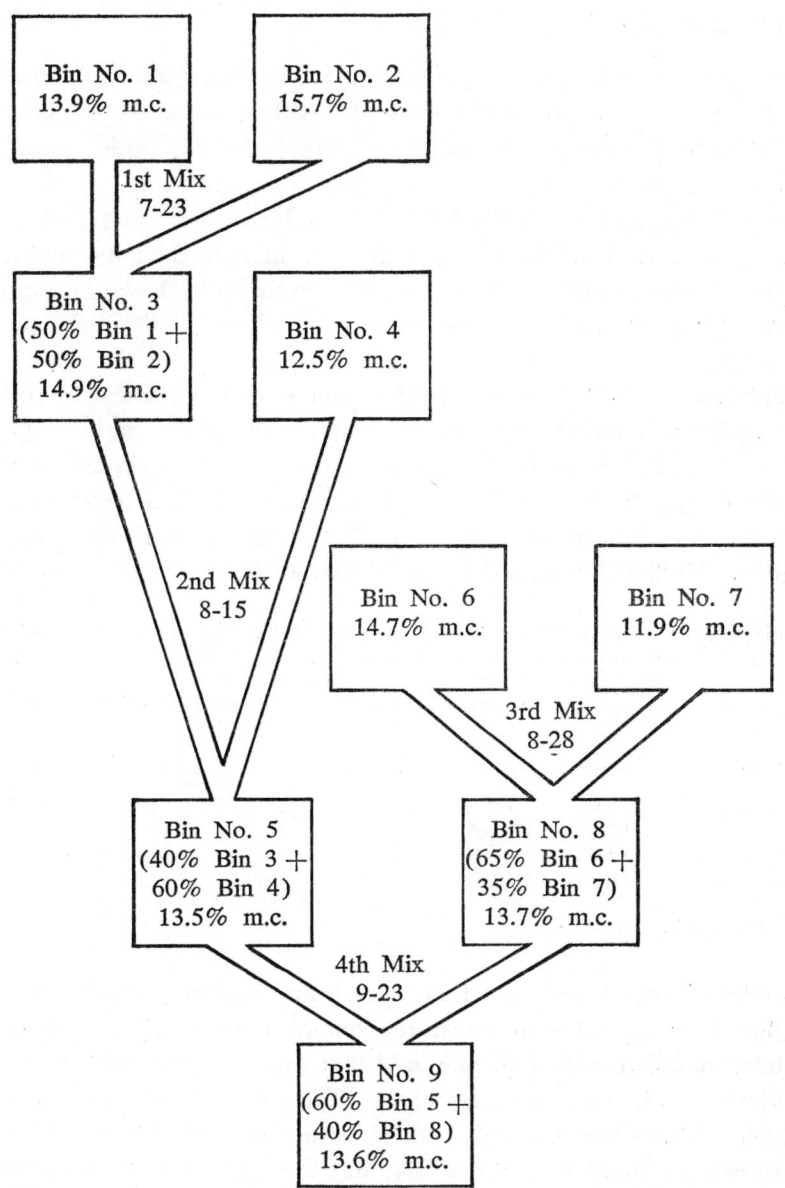

History of a series of mixes. On 10–23, 5% sick wheat was detected in Bin No. 9. Samples were withdrawn and their moisture content determined; by the Tag-Heppenstal meter the moisture contents averaged 13.4% but by oven drying they averaged 14.1% and ranged from 13.9% to 14.6%. These samples were stored in the laboratory and after two years they had developed 100% sick wheat. (The figures for moisture content given in the diagram were obtained by use of the Tag-Heppenstal meter.)

least a year. Germ damage sufficient to reduce the grade and price of all this grain developed in three months, with a loss to the elevator of about $1.00 per bushel. A number of samples of the grain from this bin were stored in the laboratory; they were gradually invaded by storage fungi, and eventually 100% of them developed brown and damaged germs. Had all the grain in this mix actually had the moisture content it was thought to have according to the data shown in the diagram, it could have been stored for a year or more without danger of spoilage.

Fairbrother (31) in 1929 mixed lots or parcels of wheat of different moisture contents, and found that the originally moister grain retained, in the mix or blend, a moisture content 1%–2% above that of the originally drier grain. The theoretical or mathematical average moisture content was never reached. Data from his work are given in Table 8. Hubbard *et al.* (32) showed that at a given relative humidity

Table 8. Original and Final Moisture Contents of Samples of Wheat Mixed Together in Equal Proportions and Allowed to Come to Equilibrium (in Percentage)

Mixture	Original Moisture Content			Final Moisture Content		
	1st Sample	2nd Sample	Average	1st Sample	2nd Sample	Average
1	17.28	12.08	14.68	15.39	13.80	14.58
2	17.33	10.07	13.70	14.79	12.71	13.75
3	18.53	9.76	14.14	15.45	13.65	14.52
4	18.41	12.68	15.55	16.80	14.40	15.60

Source: Fairbrother (31).

samples of wheat and corn that were losing moisture would retain a higher final equilibrium moisture content than samples that were gaining moisture, although they said that this hysteresis effect was not evident at relative humidities above 75%. Some of the samples with which Fairbrother worked had moisture contents in equilibrium with relative humidities above 75%, and the hysteresis effect was very evident. Schroeder and Sorenson (33) showed that the moisture content of rough rice in equilibrium with a relative humidity of 85% ranged from a low of about 15.0%, if the rice was gaining moisture, to a high of about 16.0%, if the rice was losing moisture. In evaluating storability of a given lot of grain, there is little practical value in measuring the moisture content of a given sample with extreme precision, and expressing it to the second place to the right of the decimal point, when within the representative sample itself half the kernels may have a moisture content of 13.5% and the other half a moisture content of 15.0%, and the portion of higher moisture content is already moderately invaded by storage fungi. The formidable problem that some mixtures present to grain dryers will be discussed in Chapter 8.

One very important factor that increases the storage risk of such mixes or blends, of which warehousemen are not generally aware, is this: that portion of the grain of higher moisture content going into such a mix may have had a high moisture content for weeks or months; it may already have been invaded by storage fungi sufficiently to be on the verge of developing detectable damage, and be of very high storage risk.

A common reason for mixing off a lot of poor quality grain is to conceal its poor quality. In Chapter 8 will be found a report of a USDA investigator, a pioneer in the study of aeration of stored grains. In one of his tests with wheat, about 20% of the grain in one bin became musty (91). He said, "However, after thoroughly mixing with the balance of the wheat, the odor of the musty portion was not strong enough to be detected." In other words, this lot of partially spoiled grain could be passed off as sound grain, because the buyer had no means of detecting its condition! This is an accepted practice in grain merchandising; it is even tacitly approved by the Official Grain Standards, since 5% decayed kernels are allowed in grade No. 2 corn, 4% in grade No. 2 wheat, and 7% in grade No. 3, 10% in grade No. 4, and 15% in grade No. 5 of both corn and wheat. The particular lot of wheat mentioned above, composed of 20% musty grain (plus probably at least an equal amount moderately invaded by fungi but not yet obviously musty), would constitute a very poor storage risk. Once the fungi have invaded grain, they will continue to grow in it at a lower moisture content than they otherwise would. Also once they have invaded it, the grain absorbs moisture much more readily than sound grain, and loses it more slowly than sound grain. The hysteresis effect is accentuated.

If, say, 20% of such fungus-invaded, spoilage-prone, high-moisture-content grain is mixed with 80% of sound grain of low moisture content, the most probable result is that the 20% portion will develop damage or spoilage and the whole bulk will thereby be reduced in grade. Almost unquestionably this is the course of events in many cases where, for example, up to 20% germ damage has developed in grain in a given bin within a few months after harvest, but with continued storage the percentage of germ damage has not increased. Many experienced warehousemen load their bins with mixes of grain of which a portion is already in the beginning or intermediate stages of spoilage, or at least moderately to heavily invaded by storage fungi, then attribute subsequent spoilage to "mysterious" causes.

If, in addition to this very real hazard, the meter in use by that elevator or warehouse gives readings 1% low, and the mixes or blends of grain, instead of having an average moisture content of 13.5%–14.0%, have a moisture content of 14.5%–15.0%, trouble is almost inevitable. An additional danger is that the high-moisture-content lots themselves may be the product of previous mixes involving one or more portions of even higher moisture content, of higher invasion by storage fungi, and of higher storage risk. It is far better to dry grain of high moisture content and high storage risk artificially, and segregate such lots as much as possible, rather than to mix them with lots of low moisture content and high quality in the hope that they will in this way be protected or upgraded.

It is impossible for a buyer to detect mixes of high storage risk by any method of moisture measurement now in use, although work is under way on this. However, if any of the grain has been stored at a high enough moisture content long enough to have been invaded by storage fungi, this can be detected easily by culturing the seeds in the laboratory. The practice of mixing or blending of different lots of different storage behavior to achieve a given arithmetical average moisture content is decreasing, but it probably will long continue. At least the means now are available to detect which of such mixes may constitute a high storage risk, and it is possible to sample such bulks periodically and, by laboratory tests, to determine some time in advance whether any spoilage is likely to occur.

Differences in moisture content among individual kernels or seeds. There is some evidence that the moisture content of individual kernels of corn in a supposedly uniform lot may differ by at least 0.5%. As indicated above, if the supposedly uniform lot actually is a mixture of kernels of different moisture contents, the differences may be greater than that. In one lot of corn from commercial storage we found differences in moisture content of 2.0% among individual kernels. This may partly explain why, in corn stored with an average moisture content of 14.5%, some individual kernels are invaded heavily and rapidly by storage fungi, while others are invaded only lightly or not at all. We encounter the same phenomenon—some individual kernels or seeds invaded heavily, some not invaded at all — in all types of grains and seeds that we have stored at moisture contents near the lower limit that permits invasion by a given species of fungus. It would be desirable to obtain conclusive evidence to explain why this is so.

One possibility is that individual kernels or seeds differ from one another in inherent or genetic resistance to invasion by storage fungi. There is no proof, but one would expect them to differ in this as they do in resistance or susceptibility to attack by other fungi. It might be possible to breed varieties much more resistant to attack by these fungi than are the present varieties. Many plausible and logical arguments can be advanced to show that such an approach would be without value, but this can be determined only by research, not argument.

Effect of insects and mites on moisture content. The increase in moisture content resulting from activities of insects and mites has been discussed in Chapter 2.

Importance of taking samples. These complications and qualifications relating to an accurate knowledge of moisture content in different portions of a mass of grain are not theoretical but actual; they have been discovered and elucidated by work with grains in commercial storage.

Obviously, then, obtaining accurate knowledge of moisture content requires more than a good meter. This must be combined with an alert appreciation of some of the limitations of any method of estimating or measuring moisture content of a large mass of grain on the basis of small samples removed from the mass, and with a knowledge of the agencies and conditions that contribute to storability and to spoilage. The Probe-A-Vac developed by Cargill, Inc., makes it possible to obtain, easily and quickly, a sample of any desired size from almost any kind of grain or seed stored in bulk in almost any kind of bin or warehouse. By taking samples from different portions of the bulk at intervals during the storage life of the grain, and testing them for moisture content with a method or machine that is known to give accurate measurement, the risk of unexpected or hidden spoilage can be greatly reduced. If the samples are also examined for incipient germ damage, and tested for numbers and kinds of fungi, the warehouseman at all times can have an accurate knowledge of the condition of the grain in all portions of the warehouse.

4 *HEATING AND RESPIRATION*

HEATING

The problem of heating of stored grains probably is as old as the practice of storing grains, but only very recently have we learned the causes of heating, the processes involved, and how it may be prevented. The magnitude of losses suffered from heating and accompanying spoilage in stored grains and grain products is not known, and probably would be difficult or impossible to determine. However, a couple of case histories may serve to indicate the nature and consequences of heating in stored grain.

Plate 2 illustrates the final product of heating in a commercial bin of soybeans. The bin was loaded in the fall of 1962. According to the warehouse records none of the lots of beans that went into this bin had a moisture content higher than about 13.5%. This figure of 13.5%, it must be emphasized, was the figure in the warehouse records, not necessarily the moisture content of the beans in the bin, especially after they had been there for several months. (Later in the chapter we will encounter another case of heating of soybeans, in which the beans averaged 15.5% moisture content as loaded into the bin, but when the bin was emptied only two months later some of the beans had a moisture content of 28%.) No samples were ever taken from the bin loaded in the fall of 1962; it was *assumed* that the beans in the bin had the moisture content shown on the warehouse receipts — that is, no higher than 13.5%. No consideration was given to the possibility of moisture shifting in the mass of beans during the winter. Also, only the average moisture content of each lot of beans going into the bin was determined, and it is quite possible that some portions of some lots had a moisture content at least 1% above the average figure, which would increase their storage risk. It is not unlikely that some of these lots included beans that already had been invaded to some extent by storage fungi, which also would increase the storage risk. Finally, it was assumed that beans with a moisture content of 13.5% would be safe from deterioration, an assumption that seemed reasonable in view of the Official Grain Standards of the United States, which permit a moisture content of up to 14% in beans of grade No. 2. It now is known that beans with a moisture content of 13.0% will be invaded by storage fungi within four or five months at a temperature of 75° F., but the man in charge of these beans did not know this.

Samples should have been withdrawn occasionally from this bin, tested for moisture content, and inspected for storage fungi and discoloration. However, this was not done.

In the spring of 1963 the temperature began to rise slowly in portions of the beans some distance below the surface, as indicated by readings from temperature detection cables in the bin. This slow rise continued through the summer, without causing undue concern to anyone responsible; no samples were taken to determine the extent of the hot spots, or the extent of deterioration, or whether in portions of the bulk there might be temperatures higher than those shown by the temperature detection devices — as seems highly probable. The first significant rises in temperature in May should have been taken as an indication that spoilage already had reached an advanced stage in some portion of the bulk, rather than as an indication that there might possibly be some trouble at some indefinite time in the future.

Finally in late September the temperature rose to over 200° F. in portions of the bulk, and when air was admitted by cutting a hole in the wall near the bottom of the bin, to over 400° F., which was as high a temperature as the potentiometer in use there could measure. When the mass had cooled enough to be removed — no easy task, since large portions had solidified so that they had to be broken up by force — over 90% of the mass was a black and stinking mess, and the rest caked together with fungus mycelium. The loss was over $400,000.

In Cairo, Illinois, in the mid-fifties, 242,000 bushels of winter wheat, grown and harvested in the surrounding region and brought to the elevator by truck, were stored in a large bin. The *average* moisture content of the grain was 13.2%, although according to the warehouse receipts some of the individual truckload lots had a moisture content of 14.0%.

51

Strangely enough, although many lots had moisture contents such as 13.54%, 13.73%, 13.87%, 13.99%, not a single one had a moisture content over 14.0%. A grain merchandiser is more or less forced to deal in such grain as is offered to him, and in this case it is a fair presumption that if any of the truckload lots did have a moisture content above 14.0%, and should therefore have been graded "tough" and paid for at a lower price to the farmer, the warehouseman thought that he could mix or blend them with lots of lower moisture content to achieve an average moisture content low enough for safe storage. In those lots which had an average moisture content of 14.0%, it is entirely possible and even highly probable that some of the grain had a moisture content of 15.0% or more; it was being harvested in a period when there were intermittent showers throughout the region almost every day, and it would be strange indeed if some of the grain did not have a moisture content high enough for risky storage. The moisture contents of all average samples were measured with a Tag-Heppenstal moisture meter, and in our experience it is not too unusual for this meter to give readings as much as 1% low on some samples, as compared with moisture content determined by oven drying. We do not know that the particular meter used for these lots of grain was inaccurate; neither did the warehouseman know. If the meter was giving low readings, then possibly some of the grain in some of the lots had a moisture content of as much as 16%, instead of all the grain having an average and uniform moisture content of 13.2% as the warehouseman maintained. Later evidence indicated that some of the grain almost certainly had a moisture content of 16%.

The temperature of the grain as it was brought from the fields ranged mostly between 80° and 90° F., but there were considerable differences in temperature among the loads, some being of higher, some of lower, temperature than this. With the advent of cool weather in the fall, rapid moisture transfer might be expected to occur in the bulk. No samples were ever removed from different portions of the bulk during storage for tests of moisture content or incipient spoilage.

In December the grain began to heat, at first slowly, then more rapidly. When the temperature got high enough to alarm the warehouseman (each warehouseman has his own threshold of alarm in response to temperature rises in the grain for which he is responsible) he kept his crew working day and night to transfer the grain and prevent further spoilage. The heating ceased, but when the grain was loaded out in early spring it had 40% germ damage and was rated as sample grade, with a loss to the warehouseman of $242,438.46. Typical seeds from the grain as it was loaded out of the bin are shown in Plate 3.

The warehouseman maintained, in the resulting lawsuit, that he had exercised all due care and caution in the storage of the grain, and that the spoilage was a result of something that happened to the grain before harvest. He personally was convinced that the late spring frost, which had killed back many of the wheat plants when they were about a foot high, was the major factor responsible for the poor storability of the grain harvested from those plants. Various other natural and supernatural phenomena were also suggested as possible causes. This was, however, another case of heating and spoilage of moist grain caused by storage fungi, and there is no question that the warehouseman himself was responsible for the loss.

Cases such as these are by no means rare, and in some years in some regions losses of this sort are distressingly frequent. In the entire economy of the country these may not amount to a great total, but for individual firms they may represent a serious (and needless) drain on profits, and for the small operator or the individual warehouseman they may be a tragedy.

RESPIRATION

Respiration and heating are considered together here because they are parts of the same phenomenon, parts of the same biological process which results in spoilage. Moist stored grains are not the only organic materials that heat in storage; such heating occurs in baled cotton and wool, in hay, in feeds, in manure piles and compost heaps. Manure composted for the cultivation of mushrooms is allowed, indeed encouraged, to heat to about 145° F., to make it suitable for the growth of mushroom mycelium or spawn. It has long been known and generally recognized that in many of these materials microbes are responsible for most or all of the heating up to a temperature of 70°–75° C. (158°–167° F.) — fungi at the lower moisture contents where no free water is available, bacteria at higher moisture contents. Some of this work was done before 1900 (39). It is a biological truism that all living things respire, although this is not always experimentally provable by present techniques, especially in such things as dormant seeds or dormant spores of bacteria that remain viable for more than fifty years. When hexose sugars are being consumed for food, regardless of who or what is consuming them, the formula for respiration is as follows:

$$C_6H_{12}O_6 + 6\,O_2 \rightarrow 6\,CO_2 + 6\,H_2O + 677.2 \text{ calories}$$

In a laboratory it is easier to measure the amount of carbon dioxide evolved than it is to measure the amount of heat produced. For this reason many laboratory studies aimed at evaluating the different factors involved in and contributing to spoilage of stored grains have been devoted chiefly to measurements of amounts of carbon dioxide evolved.

One of the major questions in these studies of respiration of stored grains, although not always recognized as such, was whether the respiration was due mainly to the seeds themselves or to microflora, particularly to storage fungi, growing on and in the seeds being studied. A considerable amount of work had to be done before the question could even be posed as definitely as this, and much more work — some of it requiring fairly sophisticated apparatus and techniques — had to be done before it could be answered. The answer now seems definite and clear — at moisture contents up to those necessary for germination of seeds, most of the respiration is due to fungi, storage fungi especially.

This was not at all definite and clear to some of the investigators, and there may be a few who still question the role that fungi play in the respiration and heating of moist stored grains and seeds. Some of the experimental work involved in gaining the answers to this seemingly devious but actually straightforward problem are worth examining more closely, partly to illustrate the fact that the truth often is elusive, and that the validity of conclusions depends on the validity of the tests on the results of which the conclusions are based.

In some earlier tests, moist grain was held at temperatures of 40° and 45° C., and the carbon dioxide produced was assumed to be produced by the grain, because the stated aim of the study was to measure respiration of the grain. There evidently was no realization that at a temperature of 45° C. moist grain would be killed within a few days at most, but that storage fungi would thrive mightily, although perhaps unseen by anyone not familiar with them and not looking for them. Many tests between 1918 and 1940, with a variety of grains and seeds, were at a temperature of 37.8° C., for reasons not explained. So far as can be determined, in none of these were preliminary tests made to determine whether this temperature, at the moisture contents used and within the time span of the tests, was lethal to the seeds. To measure respiration of living seeds, one must work with living seeds.

Many of the tests were with seeds in sealed containers from which the atmosphere was removed at the end of the test period, usually several days, and the amount of carbon dioxide measured. It was not realized that fungi growing vigorously on the grain might within a few hours produce enough carbon dioxide to reduce the respiration rate greatly.

Even moisture content was misinterpreted. Coleman and Fellows (40) in 1925 determined the moisture contents of various grains and seeds in equilibrium with various relative humidities. This was fundamental work, carefully done, and is still referred to with respect. Being engineers, they expressed moisture contents on a dry weight basis, which they stated only briefly and in passing in the text; to them this procedure needed no emphasis, since it was the usual and proper way of expressing moisture content — for engineers. But they published the data in *Cereal Chemistry*, most of whose readers habitually deal with moisture content expressed on a wet weight basis, which is the normal and proper way for cereal chemists to express moisture content, because this is the basis used in the grain trade.

Fifteen years later one of the authorities in the field of respiration of moist stored grains used the data of Coleman and Fellows to support the contention that "molds will not proliferate freely until the moisture content of wheat is in excess of 17.5 per cent moisture [*sic*], which is substantially above the level permitted under U.S. Grain Standards for the grades commonly accepted for delivery under contract" (35). A moisture content of 17.5% on a dry weight basis is equal to a moisture content of 14.8% on a wet weight basis.

This is cited not with any intent of impugning the work of the man who made the statement quoted, one of the world's recognized leaders in the field of cereal chemistry, but merely to illustrate an error that was made on what turned out to be a very important point indeed. Even authorities can sometimes be wrong. That Coleman and Fellows expressed moisture content on a dry weight basis instead of on the wet weight basis used by cereal chemists was evidently first detected by Milner and Geddes (48) in their work with respiration of stored soybeans in the 1940's. Some investigators are still unaware of this discrepancy.

As a final point, none of the few who scouted the idea that fungi might be mostly responsible for the respiration of moist stored seed determined the number and kinds of fungi on their samples before, during, and after the tests. Without such data their opinions on the possible role of fungi were valueless. Some of the tests, it is true, purported to exclude fungi from the seeds under study by the use of supposedly fungicidal materials, but there were no preliminary tests to determine whether the materials used, in the concentrations used, actually were fungicidal under the conditions of the tests. Some of the materials used as supposed fungicides were even known from previous work to be not especially fungicidal when applied to moist stored seeds. *Obviously, to determine how much of the measured respiration, or carbon dioxide, was produced by the seeds themselves, and how much by the fungi, required a joint attack by those familiar with the biology of grains, with the biology of fungi, with the respiratory phenomena of both, and with the techniques of differentiating the two. Essentially it meant working with seeds known to be free of fungi, not just supposed or assumed or thought or believed to be free of fungi capable of growing under the conditions of the tests.* This turned out to be difficult.

TESTS WITH FUNGICIDES

Darsie, Elliott, and Peirce (41) in 1914 determined the amount of heat produced by germinating seeds, that is, seeds of much higher moisture contents and tremendously higher rates of respiration than could ever be expected from seeds at moisture contents usual in stored grains. They were interested mainly in evaluating the germination vigor of seed lots, and thought that one approach would be to determine the rise in temperature that accompanied germination — the more vigorous seeds presumably would respire more and produce a greater rise in temperature than less vigorous seeds. They used various disinfectants or fungicides to reduce or eliminate the microflora on the outside of the seeds, since even at that early date they recognized that this was necessary if heat production by the germinating seeds themselves was to be measured accurately. They settled upon the use of mercuric chloride, which they said was injurious to some seeds and did not eliminate microflora from others, but was better for their purpose than copper sulphate or other materials they tested. This is mentioned because some later workers used either or both of these compounds to treat moist seed before storage, and then assumed (and so stated) that, because they had treated the seeds with a fungicide (presumably with a capital "F"), fungi were eliminated from the seeds and so could be eliminated from the conclusions. A false assumption, as we shall see.

The seeds so treated by Darsie, Elliott and Peirce were moistened enough to germinate, put in clean Dewar bottles which had been checked individually before use to make sure that their insulating properties were what they were supposed to be, and incubated at a temperature of 17.6° C.; the temperature of the seeds was measured twice daily. They worked with barley, oats, wheat, corn, hemp, and red clover.

Seeds of high germinability that produced vigorous seedlings increased the temperature within the bottles 1°–3° C. during the most vigorous phase of germination, when respiration presumably was highest. However, some lots were overgrown with molds, because the treatment with mercuric chloride by no means eliminated all fungi from the seeds, and in these bottles the temperature rose as much as 10° C. within four or five days, at which time the tests were terminated. Concerning this unusual rise in temperature they stated, "Upon opening the flasks, the cause of this sudden rise was at once seen to be the active fermentation, the odor of which was evident, and the contents of the flask were found to be covered with Penicillium." At least they actually looked at the seeds at the end of the tests.

Gilman and Barron (42) in 1930 reported the results of work on the effects of molds on the temperature of stored grains. Gilman was a mycologist, and knew his way around with fungi. They "sterilized" seeds with dry heat or with a solution of alcohol and mercuric chloride. Their use of the word "sterilized," is somewhat unfortunate, because the word means, in this context, "to render sterile, or free from pathogenic bacteria and other micro-organisms, as by heat or by chemicals," and so a reader not familiar with some of the limitations of these "sterilization" procedures is inclined to assume that the word means what it says, and that the seeds *were* free of micro-organisms. It would have been more accurate to say that they had steeped or shaken the seeds for such and such a time in an alcohol-mercuric chloride solution and hoped or assumed that this would eliminate or greatly reduce the growth of microflora. They present no evidence to show that they actually did eliminate microflora from the seeds, but they evidently greatly reduced the growth of fungi. Some seeds were moistened enough to germinate and put in sterile Dewar bottles; others were adjusted to 18% or 20% moisture content, inoculated with each of several species of fungi isolated from stored seeds, and similarly stored. Their results are summarized in Table 9.

Table 9. Temperature Increases (in Degrees C.) Resulting from Germination of Seeds Relatively Free of Fungi and Seeds Inoculated with Fungi

Condition	Oats	Wheat	Barley
Relatively fungi-free seeds during germination	6.8	2.6	3.6
Seeds inoculated with *Aspergillus flavus*			
18% m.c.	26.0	7.4	16.0
20% m.c.		15.8	
Seeds inoculated with *Aspergillus niger*			
18% m.c.	26.3	9.8	21.9
20% m.c.		23.4	
Seeds inoculated with *Aspergillus fumigatus* (18% m.c.)	26.4	5.2	

Source: Gilman and Barron (42).

The moist germinating seeds entirely or relatively free of fungi produced only slight increases in temperature compared with the increases in temperature resulting from the growth of fungi. The smaller increase in temperature caused by the three species of fungi on wheat at 18% moisture content is easily explained by the fact that all these species require a moisture content of more than 18% in wheat to grow at all vigorously.

Larmour *et al.* (44) in 1935 studied the respiration and heating of damp wheat. They stored wheat at moisture contents of 16% to 24%, and used carbon tetrachloride in concentrations sufficient to control the fungi but not kill the wheat. In this case, "control" does not mean elimination of the fungi, since some of the samples so treated were musty at the conclusion of the tests. They concluded that there are two kinds of respiration in moist wheat, that due to the embryo of the seed and that due to molds, and, if molds can be inhibited, wheat at 18%–20% moisture content will not respire any faster than that at 12% moisture content, and therefore will not heat.

Milner *et al.* (46) tested more than 100 compounds as possible fungicides or fungistats on wheat. They stated: "Few of these effectively inhibited the growth of molds on or in the seed. Some compounds inhibited certain molds but not others, or inhibited the surface growth and spore production of certain molds without preventing the growth of the molds in the interior of the seed. This suggests that the effectiveness of a given compound in inhibiting the development of molds on or in moist stored seed of any kind can be ascertained only by determining the number and kinds of molds originally present, and their subsequent increase or decrease after the seed has been treated with the supposed fungicide." This last sentence may seem to be a belaboring of the obvious; it was a sort of back-of-the-scientific-hand to one investigator of respiration who stubbornly insisted that fungi could not possibly be contributing to the respiration of moist stored seeds because he had applied materials *labeled* fungicides to the moist seeds, and the seeds still respired. As a student facetiously said, the storage fungi could not read the labels. There are various pitfalls in testing the fungicidal value of purported fungicides. A compound may be exceedingly toxic in agar, or in seeds with a moisture content high enough so that free water is present (in which case it is likely to kill the seed also) but not at all or only slightly toxic to fungi in seeds with a moisture content of 16%–18%. And when seeds to which a fungicide has been applied are cultured in agar to determine if fungi grow from them, it is necessary to remove or inactivate the fungicide — otherwise it may diffuse out into the agar and keep the fungi from growing on the agar although it did not keep them from growing on or in the seeds. For many years plant pathologists thought and taught that the seed treatment fungicide New Improved Ceresan (ethyl mercury phosphate, 3%) killed the fungi in seed treated with it. They had proved this by placing seeds treated with it in agar, and showing that no fungi grew from such seeds, whereas many fungi grew from untreated seeds of the same lot. What this test proved, of course, was that no fungi grew from the treated seeds put on the agar. That the fungi in the seeds were dead was an assumption and, as it turned out, an erroneous one. Moore and Olien (49) and Olien and Moore (50) treated seeds with various mercury compounds, stored the seeds for some time, then inactivated the fungicides by washing the seeds with other chemicals, and cultured the seeds in agar. Fungi grew from all of them, the same kinds of fungi and in the same numbers as grew from untreated seeds. Even some professional plant pathologists are unaware of this.

There is another complication in the use of fungicides to reduce or eliminate the growth of storage fungi. The seeds of cotton have a dense, thick, impermeable seed coat, although there is an Achilles heel in the shape of a small opening at the place of attachment, and fungi can easily grow through this opening into the interior of the seed and thrive there mightily. Fungicides applied to the outside of the seeds have no effect on these interior fungi. Yet Malowan (45) applied formalin, copper sulphate, and mercuric chloride, none of which are very effective seed treatment fungicides in the first place, to cottonseed, and claimed, "The results show that disinfecting solutions do not prevent the generation of CO_2 and that bacteria, molds or yeasts can not be the cause of it." He was as wrong as could be, and yet this totally worthless and invalid "evidence" was cited in a research paper twenty years later as proof that fungi could not be involved in respiration of moist stored seeds.

FUNGI RECOGNIZED AS MAJOR SOURCE OF RESPIRATION

Ramstad and Geddes (51) measured respiration of whole and split soybeans at different moisture contents. They stated, "The soybean respiration experiments already described provide very strong circumstantial evidence that micro-organisms are responsible for most of the respiratory activity observed in soybeans of high moisture content which have been kept at a relatively high temperature." They used a temperature of 37.8° C. because, as they noted, this was the temperature that Bailey and Gurjar had used in 1918 (36). They also said: "Several lines of evidence indicate that the high respiratory rates observed after storage at high moisture levels and moderately high temperatures were due chiefly to micro-organic activity. Beans which exhibited these high respiratory rates usually were visibly moldy or had a musty or sour odor and bacteria and a number of fungi were found in abundance. Moreover, *higher respiratory rates were observed in nonviable beans than in beans of high viability* [italics ours]." In other words, the dead beans respired more rapidly than the living ones. This was the first of a long series of papers by Geddes and his co-workers on the relation of microbes, and especially storage fungi, to the respiration and heating of moist stored seeds.

In spite of this rather clear-cut evidence, plainly stated and more than twenty-five years old, many warehousemen at the present day, responsible for hundreds of thousands of bushels of high-priced soybeans in their bins, are reluctant to believe that storage fungi are or can be of any significance in their storage problems, and only smile tolerantly, or gaze off into the distance in mild disdain, if this is suggested to them. Any experienced warehouseman knows that soybeans, like corn, have an urge to heat and germinate in the spring. They sometimes have an urge to heat and germinate in the winter, too, as Geddes and Ramstad showed in their work with soybeans in a commercial bin.

The bin had been filled in late November in Minneapolis, Minnesota, when the outdoor temperatures were relatively low. The twelve carload lots with which the bin was loaded were of grades No. 2, 3, and 4, with moisture contents ranging from 14.0% to 18.0%, as determined by the Tag-Heppenstal moisture meter, at that time the officially approved meter for determining moisture content of grains. Presumably only one average sample from each carload lot was used to determine moisture content of that lot, since that was and continues to be the customary practice. The average moisture content of all the beans was 15.5%.

To obtain records of temperatures in different portions of the bulk, and also samples of interseed air for analysis, a pipe one-half inch in diameter was forced down into the beans at the desired location, each time and place of sampling. They stated, "No readings could be taken more than 50 feet below the surface, since this was the maximum depth to which four men employing special clamps were able to force the pipe." Today, thorough sampling can be done to any depth easily and quickly with the Probe-A-Vac, described in Chapter 3.

Within 37 days, some rise in temperature occurred in almost all locations tested. After 62 days, the temperature at a depth of 40 feet was 99° F., and at 45 feet was 101° F. Because of the sour odor emanating from the beans the bin was emptied, and some 40,000 pounds of beans were found to be damaged. One sample of these beans had a moisture content of 20.3%, while a small quantity of very severely damaged beans had a moisture content of 28.0% — approximately 13% above the average moisture content of the beans, and 10% above the highest moisture content recorded when the bin was loaded. Two samples were tested for moisture content by oven drying and by the Tag-Heppenstal moisture meter, and in both of these the moisture content as determined by the meter was 1.8% to 2.0% below that determined by oven drying. Geddes and Ramstad commented: "This experiment lends further proof to the conclusion, earlier drawn from laboratory experiments, that there is a definite risk involved in storing soybeans at moisture contents permitted in the grade definitions for Nos. 3 and 4, namely 16 and 18%, as determined by the Tag-Heppenstal moisture meter." It is highly likely that some of the beans in this bin at the time it was emptied had a moisture content above 28%, since some mixing inevitably occurs as the beans run out the hopper at the bottom.

Milner and Geddes (47) measured in some detail the respiration and heating of soybeans in the laboratory, using constant aeration (which had not been done before, and which was essential to obtain an accurate measure of carbon dioxide production by rapidly respiring fungi on the beans). They also controlled the moisture content precisely and constantly, measured respiration at different temperatures from 20° to 50° C., with intervals of 5° C., and also made some attempt to determine the number and kinds of fungi on the beans at the beginning and at the end of the tests. Some of the tests lasted for fifty days. All these refinements were essential for accurate measurement of respiration and for an understanding of the processes involved; they recognized even then that fungi probably were mainly responsible for the respiration and associated changes in moist stored soybeans. If not the first to recognize this as a strong possibility, they were the first to study it in detail, precisely, and in depth. They found: "Respiration-time studies conducted up to 3 weeks yielded curves similar in form to microbiological population growth curves." They also found: "Mold proliferation, as indicated by visual condition of the seed, was positively correlated with respiratory activity (and this with degree of aeration up to the optimum) and with increases in temperature to 49° C."

In a later paper (48) they reported: "Only slight changes with time in the chemical composition of respiring soybeans as estimated by oil acid value and total and reducing sugars were noted at moisture values unfavorable to mold growth. At moisture levels where molds proliferated, drastic chemical changes occurred, approximately proportional to the moisture content and the extent of mold growth."

In a continuation of their fundamental attack on the problems of respiration they studied the influence of degree of contamination by spores of storage fungi on respiration, and attempted to eliminate these fungi by surface disinfectants or fungicides (48). They reported:

The results of the previous trial showed that even the slight micro-floral contamination of untreated normal soybean seeds yielded as rapid mold proliferation, respiration, and heating, as did heavy inocula of microflora. In an attempt to inhibit the growth of the normal seed respiration and heating in the absence of that due to the molds, a lot of the Wisconsin Manchu soybeans were "sterilized" by shaking in a solution of mercuric chloride (1–1000) for one-half minute. After decantation the seeds were washed for one minute with sterile water and allowed to drain. After 14 hours in a sealed jar the moisture content of the seed was 19.7% (equilibrium relative humidity, 87%). A 250-g sample was transferred to a sterile Dewar flask and smaller portions were put into sterile 8-ounce bottles for insertion into the adiabatic respirometers for a respiration and heating trial. Another portion of the same "sterilized," conditioned seed was heavily inoculated with spores of a pure culture of A. *flavus* and the respiratory and heating characteristics studied in the other adiabatic respirometer. . . . The surface-sterilized and surface-sterilized-reinoculated seeds heated and respired in a virtually identical manner. As the trial proceeded, both seed lots proved to be contaminated with A. *flavus* and A. *glaucus*, the percentage of seeds infected being similar for both at similar time periods. It was apparent *that surface sterilization by the technique used was ineffective against mold contaminants, which apparently existed well within the seed coats and were unaffected by sterilization at the surface* [italics ours]. This result is in complete agreement with the observation of Mead (1941) who found that seeds treated with mercurial dusts or formalin to kill surface microfloral contaminants would subsequently yield growths of mold, arising from the more deep-seated infections, when cultured on moist media.

This is important because, as emphasized before, some workers both before and after this article appeared treated seeds with various disinfectants, including mercuric chloride, and then concluded that respiration or chemical changes which occurred during subsequent storage could not possibly be due to molds. Such erroneous conclusions have been and still are accepted as completely valid by some who are innocent of any knowledge of microbiology.

In conclusion Milner and Geddes stated: "Spontaneous heating and associated respiratory characteristics of soybeans at humidity levels favorable to mold growth but unfavorable to bacterial proliferation were studied in an apparatus which maintained continuously controlled

63

adiabatic and aeration conditions, over time intervals up to 37 days. *An initial temperature increase to 50–55° C and parallel respiratory increase were directly associated with the proliferation of the molds Aspergillus glaucus and A. flavus* [italics ours]."

The conclusions of Milner and Geddes concerning the role of microflora, especially storage fungi, in the respiration and heating of soybeans were confirmed and supported by Carter (37) in experiments with wheat. He concluded, "Heating curves obtained in these experiments do not support the two-phase hypothesis of Cohn . . . but strongly suggest the conclusion that the growth of micro-organisms is the main, if not the sole, cause of heat production in stored moist wheat." He went on to say, "This further indicated that respiration in the wheat kernel is unimportant in producing the heat that developed in improperly stored wheat." And, "The data indicate that the heating of moist wheat in storage can be entirely accounted for by the energy released in the respiration of the fungi present on and in the wheat kernels."

Further evidence to indicate the predominant role of storage fungi in the respiration and associated processes of moist stored grains was provided by Hummel *et al.* (43). They stored wheat known by actual test to be free of storage fungi, at moisture contents of 15% to 31%, at 35° C., with constant aeration, and measured respiration over a period of nineteen days. They concluded: "The respiratory rates of mold-free wheat at 35° C and moisture levels ranging from 15 to 31% were low and constant with time. In contrast, the respiration of the moldy wheat markedly increased after a few days."

5 GERMINABILITY, DISCOLORATION, AND FAT ACIDITY VALUES

High germinability is required in seeds to be used for planting, for malting, and for the production of edible sprouts; not only must the percentage be high but the germination also should be rapid, uniform, and vigorous. In grains or seeds used for flour, starch, oil, or other products, germinability usually is of little moment, and does not enter into grading, although for wet milling corn of high germinability is preferred. Seeds that are obviously discolored or that have discolored germs are rated by inspectors as damaged, and so discoloration of either the embryo or the entire seed or kernel is a factor in grading. Fat acidity value has been suggested as a good indicator of condition or storability of seeds but, like germinability, it has not entered into grading.

Until the 1950's it was not generally realized that storage fungi could cause reduction in germinability of seeds, could cause discoloration of the embryos or of whole grains, and also could cause increases in fat acidity. It was thought that these changes were products of the seeds themselves, or of mysterious and unspecified causes. The evidence now is that all these changes can be produced by storage fungi, and that under the conditions likely to prevail in commercial storage, fungi are the chief causes of loss in germinability, discoloration, and increase in fat acidity values. The present chapter summarizes this evidence.

LOSS OF GERMINABILITY

There is a considerable body of evidence to indicate that invasion of seeds by storage fungi can drastically reduce germinability of the seed. (See Plate 6.) Qasem and Christensen (21) adjusted corn free of fungi to different moisture contents, inoculated some samples with various storage fungi, left others uninoculated, stored the samples at several temperatures, and periodically tested them for germinability and germ discoloration. The samples inoculated with *Aspergillus candidus* and stored at 18% moisture and 25° C. for 4.5 months germinated 9%, and most of the germs were dark brown, whereas the noninoculated controls stored under the same conditions and for the same length of time germinated 94% and had no dark germs.

Papavizas and Christensen (61) made similar tests with wheat of different classes. White wheat inoculated with *A. candidus* and kept for three months at 25° C. and 16.0%–16.4% moisture content germinated 6%, and 79% of the kernels had dark germs, whereas the non-inoculated control kept under the same conditions and for the same length of time germinated 95% and none of the kernels had dark germs. Under at least some conditions *A. candidus* evidently is a rather aggressive invader and killer of the embryos or germs of seeds. In all these tests with both corn and wheat, invasion of the germs by the fungus preceded death of the germs, and death of the germs preceded development of dark color. This is the normal course of events in almost all plant tissues decayed by fungi: first comes invasion, then death of the invaded tissues, followed by discoloration.

Tuite and Christensen (26) stored samples of barley seed at 19.4% moisture content and at room temperature, some of the samples free of storage fungi and others inoculated with storage fungi. After fifteen days the sample inoculated with storage fungi germinated 72%, whereas the sample free of storage fungi germinated 98%. They stated, "All of these species of the *A. glaucus* group, as well as *A. candidus* and *Penicillium*, invade various parts of the seed, including the germ, and directly cause or contribute to reduction in germination."

López and Christensen (58) stored corn at 19%–20% moisture content and 20°–25° C., some samples free of fungi, others inoculated with different isolates of *A. flavus*. After 74 days, the samples free of fungi averaged 97% germination, and those inoculated with *A. flavus* averaged 13%.

Fields and King (56) stored pea seeds inoculated with storage fungi, and seeds free of storage fungi, at relative humidities of 85% and 92%, and periodically determined germination percentage of the seed. At 85% relative humidity and 30° C. (86° F.) the peas free of storage fungi retained a germination of 97% for six months, whereas those inoculated with *A. flavus* decreased to zero germination in three months, those inoculated with A. *candidus* and A. *ruber* decreased to zero germination in six months, and those inoculated with A. *restrictus* decreased to zero germination in eight months. They stated, "As the uninoculated control remained free of storage molds and no reduction in germination occurred during the storage period, the reduction in germination of the inoculated seed can be attributed directly to the pathogenic action of the storage molds." Also, "Reduction in germination of the seed evidently was due to storage fungi, not to some metabolic process of the seed itself." And they concluded, "invasion by storage fungi preceded seed deterioration."

Some of the evidence from the papers cited above is summarized in Table 10. There is no question that invasion of seeds by storage fungi can cause drastic reduction in germinability. Under conditions that prevail in grain bins, storage fungi must in many cases be the primary cause of loss of germinability of the seeds.

Table 10. Reduction in Germination Percentage Caused by Storage Fungi

| Kind of Seed | Conditions of Storage | | | Germination | |
	Moisture Content (Wet Weight)	Temperature (Degrees C.)	Time	Seeds Free of Storage Fungi	Seeds Inoculated with Storage Fungi
Wheat					
Sample 1	15.5%–15.7%	20°–25°	6 wks.	95%	5%
Sample 2	16.0%–16.4%	25°	2 mos.	90%	25%–40%
Sample 3	17.0%–17.2%	25°	1 mo.	93%	20%–34%
Corn					
Sample 1	17.0%–18.0%	15°	2 yrs.	96%	0%
Sample 2	18.5%	20°	3 mos.	96%	56%
Sorghum	15.8%	28°	40 days	95%	35%
Peas	17.0%	30°	3–8 mos.	97%	0%*

Source: Adapted from Christensen and López (54).
* Fields and King (56).

That invasion by storage fungi is a direct cause of loss of germinability of seed has, for several reasons, been a difficult point to establish. First, of course, it was necessary to obtain seeds totally free of storage fungi. With some kinds of seeds, such as peas, beans, and soybeans, which are borne in pods, this has been relatively simple. If cobs of corn are selected that are free from any obvious fungus infection, the kernels are likely to be free of storage fungi. It has been more difficult to obtain even small lots of wheat or barley free of storage fungi. Some attempts have been made, in tests dealing with loss of germinability of seeds, to eliminate storage fungi by applying various fungicides to the seeds. In general these have not been very successful, primarily because most fungicides are not active at relative humidities of 75% to 85%, the range in which storage fungi thrive, and those which are active under such conditions, by virtue of toxic vapors, kill the seeds before they kill the fungi.

Another complicating factor is temperature. With increasing moisture content most kinds of seeds become increasingly sensitive to injury or death by high temperature. Hummel *et al.* (43) stored mold-free and mold-inoculated wheat at various moisture contents and at 35° C., and while they found that the major biochemical changes they studied were caused by storage fungi, even the fungus-free seeds at moisture contents of 18% and above died relatively rapidly. At those moisture contents the temperature itself was lethal. At temperatures of 20° C. (68° F.) and below, the lethal effects of storage fungi growing in seeds are relatively slow. So at one temperature the storage fungi kill seeds very slowly, whereas a temperature 10°–15° C. above that may be lethal in itself.

Germination percentage and germination vigor are affected also by a number of other factors — the conditions at harvest, seed maturity, seed dormancy, mechanical damage to the seed at harvest or in later handling, age of the seed (some kinds of seed lose viability within a few months, others live for decades, others for centuries). Seeds of soybeans die rather rapidly if stored with a moisture content just below that which will permit storage fungi to grow, and at 25° C. (77° F.), but we do not know why.

For all these reasons, plus perhaps some others as yet undetected, the problem of determining the cause of death of a given lot of seed stored at a given temperature and moisture content is difficult. Sometimes this is of much more than theoretical interest, as in the case of loss of germinability of sorghum seed mentioned in Chapter 1. If the fumigant killed the seeds, those who applied it might have to pay the losses; if fungi killed the seeds, those who stored it presumably were responsible.

Some seeds will live a surprisingly long time at rather high moisture contents and moderate temperatures, if kept free of storage fungi. Christensen (38) found that corn free of fungi and with a moisture content of 15.7%, stored for 156 days at 20°–25° C., still germinated 100%; wheat free of fungi and with a moisture content of 15.3% germinated 100% after being held for 150 days at 20°–25° C.; and barley free of fungi and with a moisture content of 15.2% germinated 93% after being held for 161 days at 20°–25° C. The physiological activities of the seeds themselves must have been at an extremely low ebb, otherwise the germinability would have decreased. Yet one finds statements in research papers dealing with grain storage, even within the past decade, to the effect that corn stored with a moisture content of 12%–15% will respire so fast that it will heat — which is nonsense. There is absolutely no evidence that cereal grains will respire measurably at moisture contents below 18%, wet weight basis. The storage fungi on them may respire very rapidly.

DISCOLORATION

That storage fungi can cause discoloration of the germs of seeds or of whole seeds or kernels is now generally recognized, although this was not true in the early 1950's. Wheat in storage sometimes develops dark germs, and in the grain trade these have been called "sick" wheat, a poor term, since such seeds are dead, not sick. Until the early 1950's no one knew the cause of sick wheat, although there were a number of theories as to possible causes — some of them bordering on the fantastic. Had anyone familiar with fungi examined embryos of sick or damaged seeds with a microscope, he would have suspected that fungi were involved, since fungus mycelium usually is abundant on such discolored embryos, and sometimes they are covered with mycelium and spores. Wheat with more than a small percentage of sick or germ-damaged kernels is likely to have a musty odor, an obvious indication that fungi might be implicated.

Christensen (52) obtained samples of wheat with 5% to 55% sick or germ-damaged kernels from the federal grain inspection office in Chicago, where they were encountered in routine inspection of commercially stored lots of wheat. The pericarps covering the germs were removed from 500 kernels from each of 26 lots, the germs examined for fungi and discoloration with the aid of a microscope, and the samples tested for germination percentage and for number and kinds of fungi. (See Plate 4.)

Germination of the sick seeds from all samples was zero — not a single damaged or sick germ was alive. In dilution cultures the germ-damaged kernels yielded an average of 402,000 colonies of storage fungi per gram — about thirty kernels — which is an exceedingly heavy load of storage fungi. The surface of some of the damaged germs was covered with a dense fuzz of fungus mycelium and spores, not identifiable as such by the naked eye but easily identifiable under the microscope. Many of the germs were so decayed that they were crumbly.

Supposedly sound seeds, with germs not discolored, were selected from each of these samples, and similarly tested. The germination of these ranged from zero to 92%, and averaged 43%, and they yielded 32,000 colonies of storage fungi per gram of grain. Really sound wheat obtained from commercial bins in Enid, Oklahoma, at about the same time, and similarly tested, germinated over 90% and yielded fewer than 1000 colonies of storage fungi per gram of grain. The damaged kernels, in other words, had more than 400 times as many living spores of storage fungi as the really sound wheat from Oklahoma, and the supposedly sound seeds from the Chicago samples had 32 times as many living spores of storage fungi as the really sound wheat. The supposedly sound seeds from Chicago were at least moderately invaded by storage fungi, the germs of many of them were dead, and the seeds were partly deteriorated, although this was not evident from inspection with the unaided eye. Obviously, if sound seeds are those not sufficiently invaded by storage fungi to be dark in color, there are different degrees of soundness. Grain moderately invaded by storage fungi and of low germinability is a much poorer storage risk than is really sound grain. Such invasion can be readily detected in the laboratory.

In several samples from Chicago with 8%–12% sick or germ-damaged seed, the average germination was above 75%, indicating that sick seeds may be found in lots with relatively high germination. Almost certainly these lots were mixes of a small amount of grain of low germinability and probably high invasion by storage fungi with a larger amount of sound grain of high germinability and low invasion by storage fungi. This may have upgraded the grain of low quality that went into the mix, but it downgraded the grain of high quality.

Typical sick or germ-damaged wheat has been produced repeatedly in the laboratory by storing samples at moisture contents and temperatures that permit storage fungi to grow — the same conditions under

PLATE 1. Storage fungi growing from the germs of (*left to right*) corn, rice, wheat, and barley. The presence of these fungi indicates poor storage.

PLATE 2. Soybeans from a bin in Kansas after a year's storage, during which no tests of moisture content were made. *Left*, a mass of beans matted together with fungus mycelium; *right*, masses of beans heated and totally spoiled, typical of the condition of the beans throughout most of the bin.

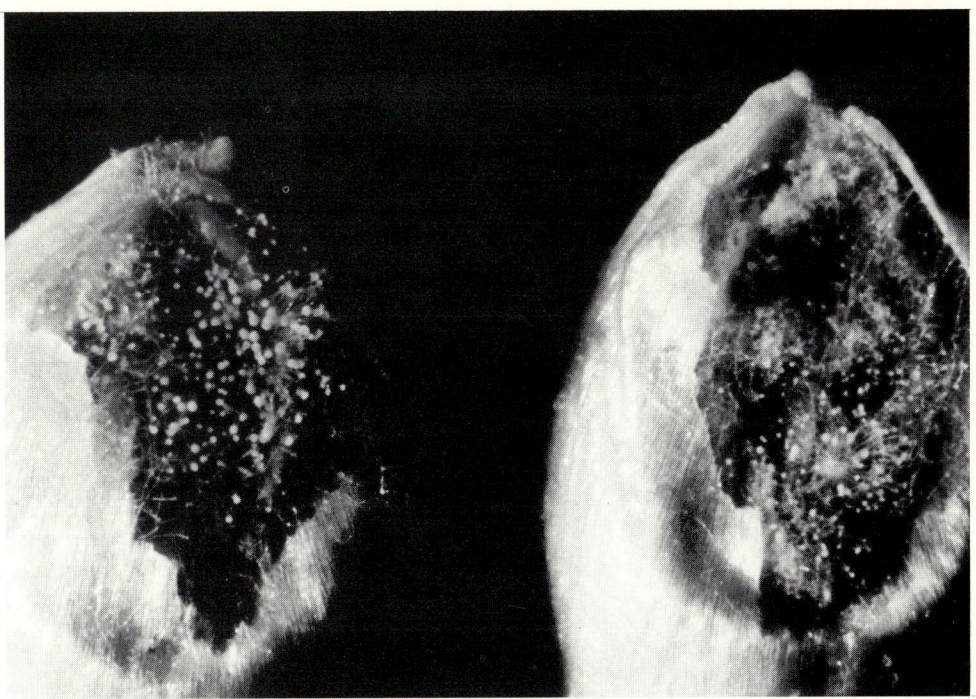

PLATE 3. Wheat kernels from a bin in Illinois, the deterioration of which was attributed to mysterious causes. The kernels were surface-disinfected, the pericarps covering the germs were removed, and the kernels were placed on an agar medium. Within 48 hours storage fungi, as shown, had grown out.

PLATE 4. Pericarps removed from the germs of wheat kernels. *Left*, sound germ; *right*, damaged or sick germ, thoroughly decayed by fungi.

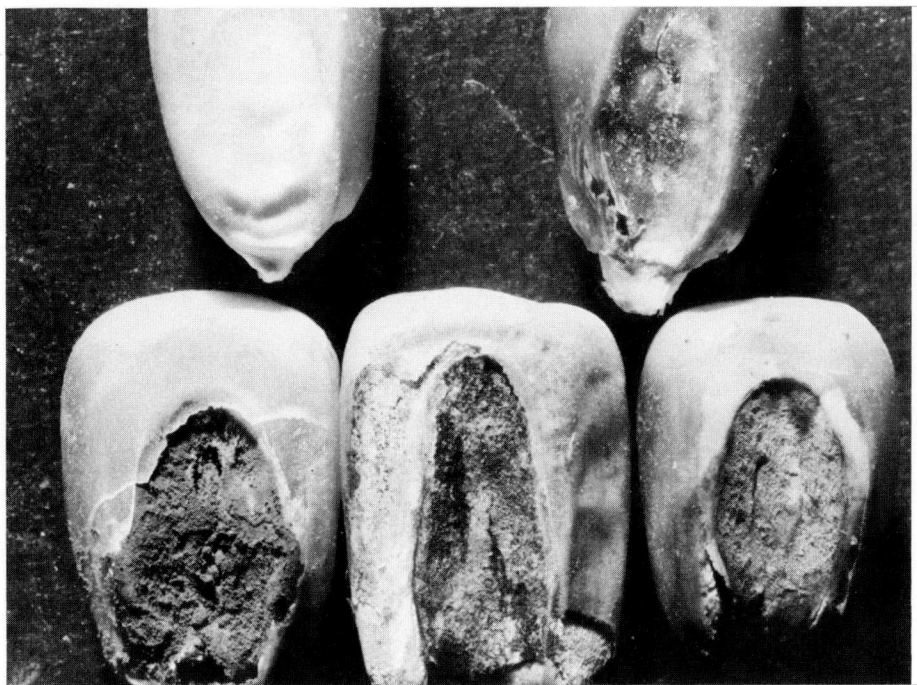

PLATE 5. *Top left*, a sound kernel of corn; *top right*, a damaged kernel with the pericarp covering the germ; *bottom row*, damaged kernels with the pericarps removed to show masses of fungus spores and mycelium covering the decayed germs.

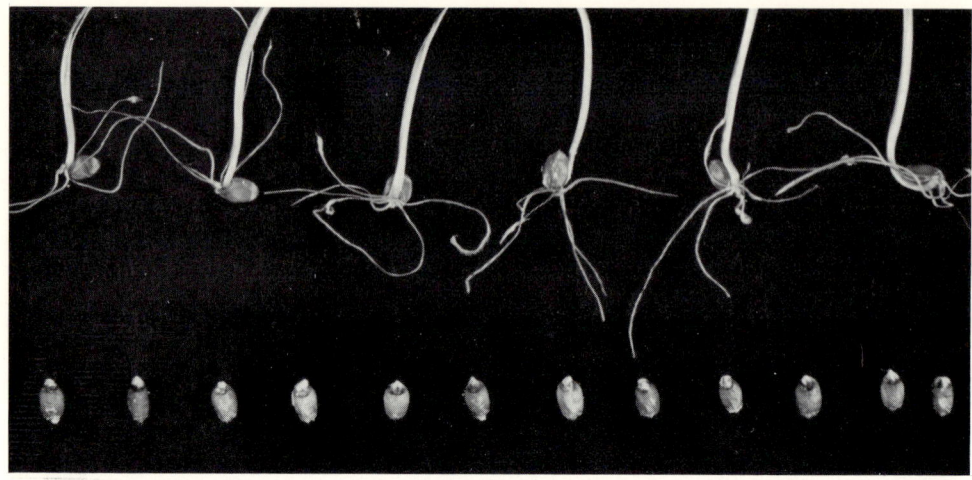

Plate 6. Reduction of germinability caused by the storage fungus *Aspergillus candidus* in wheat stored at 25° C. for 25 days with 15.4% moisture content. *Top,* in controls, not inoculated with *A. candidus,* germination was 100%; *bottom,* 100% of seeds inoculated with *A. candidus* were dead.

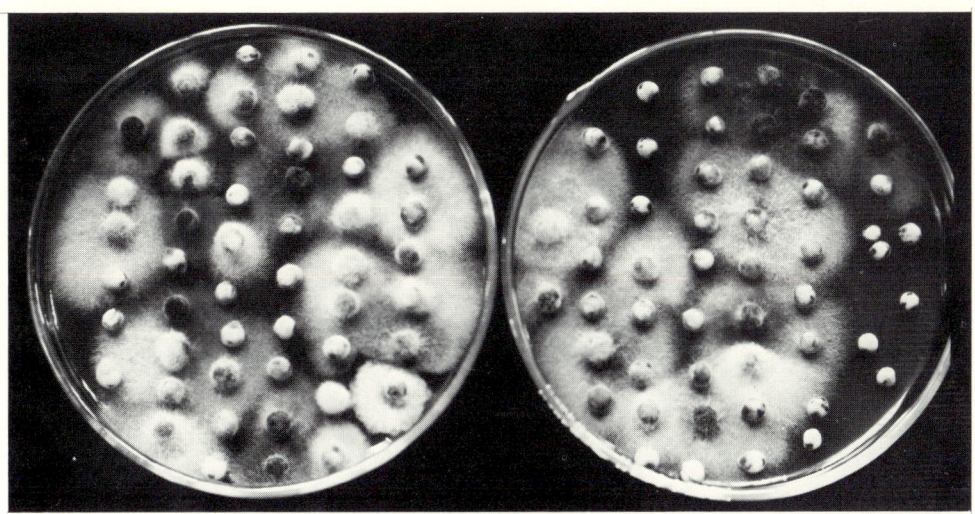

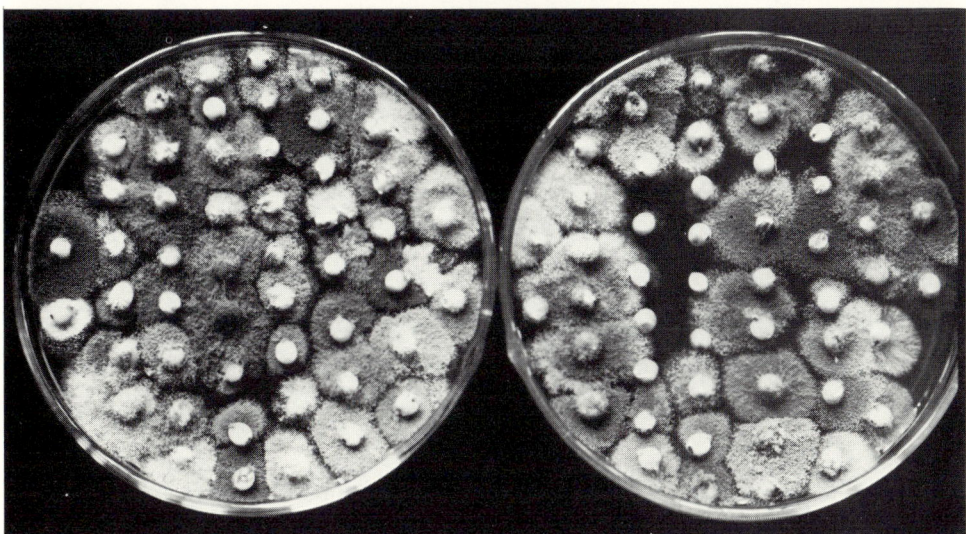

PLATE 7. Sorghum seeds from commercial storage. *Top*, in samples from six-foot depth in the bin, germination was 95%; field fungi (mainly *Alternaria*) were growing from many of the seeds but there were no storage fungi. *Bottom*, in samples from the surface of the bin, where the moisture content increased during the winter, germination was 5%–10%, with storage fungi (species of *Aspergillus*) growing from nearly all surface-disinfected seeds, indicating that the moisture content during storage must have been in the range 16%–17% for several weeks to several months.

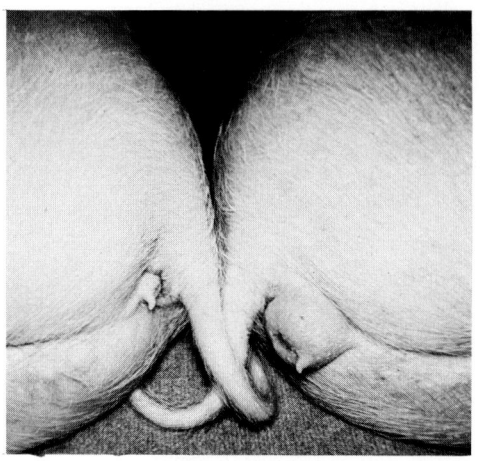 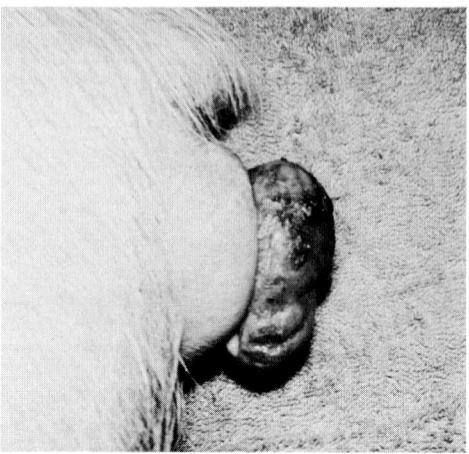

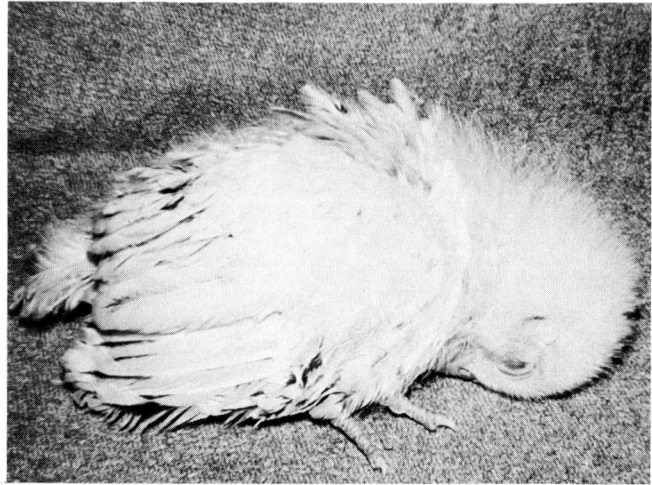

PLATE 8. Vulvar hypertrophy in gilts, resulting from ingestion of feed invaded by *Fusarium*. *Left*, normal vulva of a control gilt; *right*, swollen vulva of a gilt fed for five days on *Fusarium*-invaded feed. (Photo by G. H. Nelson.)

PLATE 9. Prolapsed vagina of gilt that had eaten *Fusarium*-invaded feed for eighteen days. (Photo by G. H. Nelson.)

PLATE 10. A chick that died from aflatoxin poisoning. For five days the chick had consumed a commercially produced feed that later was found to contain aflatoxin. The feed evidently had been made in part from grain heavily invaded by *Aspergillus flavus*. (Photo by G. H. Nelson.)

which sick wheat develops in commercial warehouses. Papavizas and Christensen (20) stored samples of white wheat for a year with a moisture content of 16.0%–16.5% and at 5° and 10° C. (41° and 50° F.) Some of the samples were free or almost free of storage fungi, and others were inoculated with various members of the *Aspergillus glaucus* group, which are common in cereal grains stored with moisture contents of 14%–16%.

The grain stored free of fungi and at 5° C. for one year germinated 94% and had no brown germs, whereas the samples inoculated with *A. glaucus* ranged from 52% to 65% in germination, and 21% to 41% of the seeds had brown germs. In the samples stored at 10° C. for a year, those originally almost free of fungi had become invaded by *A. glaucus* to some extent, but only 4% had brown germs, whereas those inoculated with *A. glaucus* had 28%–58% brown germs.

The same authors (61) stored durum, hard red spring, hard red winter, and white wheats at moisture contents of about 15%, 16%, and 17%, free or almost free of storage fungi and inoculated with storage fungi. The seeds were examined at various intervals and the number of dark germs was recorded.

After one month at 25° C. the samples with 17% moisture and almost free of storage fungi had no dark germs, whereas those inoculated with storage fungi had up to 75% dark germs. In those stored with 16% moisture and at 25° C. after two months the samples free of fungi had no dark germs, and those inoculated with storage fungi had up to 70% dark germs. White wheat inoculated with *A. candidus* and stored for three months at 25° C. with a moisture content of 16% had 79% dark germs, while the control, stored under the same conditions but free of storage fungi had no dark germs.

Christensen and Linko (53) obtained samples of hard red winter wheat from commercial bins, and stored these in the laboratory with the same moisture contents as the samples had when received. The relationship they found between moisture content, brown or damaged germs, and colonies of *Aspergillus* per gram of grain is summarized in Table 11.

Similar evidence, from work with corn, already has been presented in the section dealing with loss of germinability. All the evidence is consistent; under a considerable range of conditions spanning those likely to be encountered in practice the common storage fungi are a primary cause of germ damage.

We and others have induced the germs of wheat to turn brown without invasion by storage fungi. If wheat with a moisture content of 13% or above is kept long enough at a temperature of 35°–40° C. (95°–104° F.) the germs will turn brown even in the absence of storage fungi. The higher the moisture content and the higher the temperature, the shorter the time necessary for the brown color to develop. It is possible that,

Table 11. Influence of Moisture Content upon Increase in Storage Fungi and Percentage of Brown Germs in Hard Red Winter Wheat Stored Fourteen Months at 20°–25° C. (68°–77° F.)

Moisture Content (Wet Weight)	Brown Germs (in Percentage)	Colonies of *Aspergillus* per Gram of Grain (in Thousands)
12.3%	0	0
13.2%	0	0
13.8%	3	120
14.2%	5	330
14.7%	20	220
14.9%	28	530
15.1%	100	1008

Source: Adapted from Christensen and Linko (53).

in regions with high temperatures at harvest, grains might go into storage with a temperature of 100° F. or higher, and remain at that temperature for months, unless cooled by aeration. Under such circumstances germ damage presumably could develop in the absence of storage fungi. However, neither in the laboratory of the Department of Plant Pathology of the University of Minnesota, where since about 1948 many thousands of samples of wheat have been examined, including many hundreds with different amounts of germ damage, nor at the Grain Research Laboratory of Cargill, Inc., where since 1952 many thousands of samples of wheat from commercial storage throughout the United States have been examined, have we ever encountered a single case of germ-damaged or sick wheat in which storage fungi were not involved.

FATTY ACIDS

Among the intermediate products of spoilage in materials that contain fats or oils are fatty acids, and the characteristic odors and flavors of these fatty acids are what make partially spoiled fats rancid. It has long been known that deterioration of stored grains is accompanied by an increase in fatty acids (now designated fat acidity value, or FAV), and it has been suggested that FAV be used as a measure of deterioration, or of condition and storability of grains (86). More about that shortly.

Some of the early investigators of deterioration of stored grains believed that the lipases which converted the fats into fatty acids were produced by the seed itself. Again, as with respiration, they evidently did not question why, if a given kind of seed, such as wheat, needed approximately 25%–30% moisture to germinate, its lipases would become active at 15%–16% moisture content, whereas the fungi which were growing vigorously on the grain with a moisture content of 15%–16% would not produce lipases. That is, a dormant seed was believed to digest more fats than a rapidly growing fungus — an anomalous situation to say the least.

Among the first if not the first to attribute increases in fatty acids and associated biochemical changes in deteriorating seeds to fungi were Milner and Geddes (48), in their work with soybeans. They stated: "Only slight changes with time in the chemical composition of respiring soybeans as estimated by oil acid value and total and reducing sugars were noted at moisture values unfavorable to mold growth. At moisture levels where molds proliferated, drastic chemical changes occurred, approximately proportional to the moisture content and the extent of mold growth." At that time they (as well as we) supposed that storage fungi could not grow in soybeans with a moisture content below 14%, and so any changes that occurred in the beans stored with a moisture content below 14% were attributed not to the fungi but to processes inherent in the beans themselves. It is now known that storage fungi can invade soybeans with a moisture content below 13.0%, and so some of the older work on this point probably should be repeated.

Milner *et al.* (59), who worked with wheat, wrote: "Increased fat acidity and loss of germination accompany mold growth. On the other hand, where mold growth was virtually inhibited (nitrogen atmosphere), fat acidity and germination remained essentially constant, in spite of the elevated moisture value used in this trial."

A considerable amount of work has been done with development of fatty acids in stored cottonseeds, partly because rancidity is an important quality factor in cottonseed oil intended for use in foods. Some of the older workers contended that microflora were not involved in the

development of rancidity in cottonseed, although they had no good experimental evidence to support this contention. Christensen *et al.* (55)
worked on this problem and stated: "In general, the increase in mold
population of the stored seeds was correlated closely with increased
production of carbon dioxide and free fatty acids. Because of the peculiar structure of the seed coat, molds gain easy access to the interior of
moist stored cottonseed where they grow vigorously and sporulate pro-
fusely, so that cottonseed may be extremely moldy without betraying
any external evidence. Fungicides applied to the outside of the seed did
not prevent rapid growth and sporulation of the molds within the seed
coat." As mentioned in Chapter 4, Malowan (45) treated moist cottonseed with various supposed fungicides, and since respiration continued at a rapid rate he concluded that microflora could have played
no part in it. The peculiar structure of the cottonseed helps to explain
why his treatments were ineffective (in addition to the fact that the
fungicides he used were, for his purpose, not much good in the first
place). A fair comparison would be the application of a supposed fungicidal paint to the outside of a house to prevent growth of molds within
the house; it just would not be effective.

Nagel and Semeniuk (60) inoculated corn with pure cultures of nine
species of fungi, separately and together. The individual species differed
greatly in amount of fatty acids produced — from a minimum of 74.8
units by *Aspergillus candidus* after four weeks to a maximum of
384 units by *A. amstelodami* after two weeks. (Under conditions that
permit rapid development in stored grain, by the way, *A. candidus* will
do much more damage in a given time than will *A. amstelodami.*) With
several of the fungi the fatty acid content of the corn was less after four
weeks than after two weeks; the corn inoculated with *A. niger* had a fat
acidity value of 180.5 after two weeks and 129.0 after four weeks, a decrease of about one third. The corn invaded by *A. amstelodami* had a
fat acidity value of 384.1 after two weeks and 247.9 after four weeks,
a decrease of more than one third.

Goodman and Christensen (57) grew cultures of four species of fungi
isolated from moldy corn — *Penicillium solitum, Aspergillus flavus, A.
candidus,* and *A. amstelodami* — on cornmeal from which the oil had
been extracted, and on similar meal to which the oil originally extracted
had again been added. All caused an initial increase in fatty acids in
the meal containing oil, but after having reached a peak the fat acidity

value decreased. The amount of fatty acids produced varied with the species of fungus, and none of the fungi caused any increase in fatty acids in the meal free of oil. In these tests the fatty acids were produced as a result of fungus lipases acting upon the corn oil. The fungi tested were able to metabolize various fatty acids, and to subsist on fatty acids as their sole source of carbon. It also appeared that to some extent the lipases produced by these fungi were adaptive, since three of the isolates produced much more lipase when grown on a medium containing corn oil than when grown on a medium containing sucrose as the carbohydrate source.

This explains in part why the use of fatty acid figures as an indicator of condition or degree of deterioration of a given lot of stored grain encounters some difficulties and complications. The amount of fatty acid produced varies with the species of fungus, and very probably with strains within a species. And a given fungus may produce a relatively large quantity of fatty acids, then consume portions of them.

Also, in the work by Nagel and Semeniuk, cited above, a moisture content of 32% was used in the corn, much higher than would normally be encountered in storage. Much deterioration in stored corn and wheat occurs at a moisture content of 14.5%–16%, and at the lower moisture contents lipase production might be much less. That this is so is suggested by the results of Christensen and Linko (53). One of the samples with which they worked had, after storage for fourteen months, 5% brown germs, 330,000 colonies of *Aspergillus restrictus* per gram of grain, and a FAV of 31.42, and another sample, with 100% brown germs and 1,000,000 colonies of *A. restrictus* per gram of grain had a FAV of 32.03. Both of these samples had undergone some deterioration, the one with 100% brown germs much more than the one with 5% brown germs, yet they had essentially the same FAV, and, moreover, a FAV not appreciably higher than sound seeds.

6 *MYCOTOXINS AND GRAIN QUALITY*

Mycotoxicoses are diseases of animals (including man), caused by consumption of foods made toxic by fungi. These diseases undoubtedly have always been with us, but only recently have they come to be recognized as potentially important in the health of man and of his domestic animals. In 1882, nearly a century ago, a paper was published describing a disease of horses suspected of being caused by consumption of moldy feed (70), but little attention was paid to it. At that time probably few veterinarians even knew that there was a field of study such as mycology, and certainly no mycologist had the remotest idea that fungi might be important in the health of animals.

In 1934 veterinarians in Illinois estimated that, in the central part of that state alone, in the winter of 1933–34, 5000 horses died of "moldy corn disease" (67). Sippel *et al.* (73) reported the poisoning of 1000 swine in the southwestern United States in 1952, and of 533 in 1953, with a mortality rate of 22%; they attributed the poisoning to consumption of moldy corn, and mentioned that only a fraction of the total number of diseased animals were referred to their laboratory.

Burnside *et al.* (63), of the same team of investigators, in 1957 published the results of a follow-up investigation of this trouble. They isolated thirteen cultures of fungi from the moldy grain suspected of being toxic, of which nine were *Aspergillus flavus* and one was *Penicillium rubrum.* All the fungi were grown separately in autoclaved moist corn and the corn was fed to swine. The swine fed corn invaded by one isolate of *A. flavus* and those fed corn invaded by *P. rubrum* died within a few days, with external and internal symptoms similar to those found in the diseased animals earlier. The corn invaded by *P. rubrum* was so toxic that a mixture of only 20% of the moldy grain and 80% sound corn re-

76

sulted in the death of both 55- to 60-pound pigs to which it was fed, in less than 32 hours, after each pig had consumed less than eight ounces of the moldy feed. It also killed a goat to which it was fed. These tests were well replicated in time, and had adequate controls. Lesions in the dead animals were described in detail. The investigators soaked some of the *P. rubrum*-invaded corn in water, then administered the supernatant water to horses. Two gallons of the water extract from 4 kilograms (about 9 pounds) of the moldy corn resulted in the death, within 34 hours, of a 790-pound horse.

This work was well done, by competent investigators, and was published in the *American Journal of Veterinary Research*, a periodical of high repute and widely distributed in the United States and abroad. It caused scarcely a ripple of interest — in fact, one seldom sees it referred to in publications on mycotoxins and mycotoxicoses. Forgacs (66), the microbiologist on this team, later referred to mycotoxicoses as "the neglected diseases."

They were not to be neglected for long. In 1960, 100,000 turkey poults died in England, of what then were unknown causes. It was suspected that a new virus disease might be responsible, but investigation ruled this out. Pathogenic bacteria also were suspected, but again investigation proved that they were not involved. Eventually it was found that the common factor in the disease in all flocks where it appeared was a lot of feed produced by Oil Cake Mills, Ltd., and this firm had to reimburse the growers for the losses, to the tune of about $500,000. Not unnaturally, this stimulated research on the problem, and by 1962 the cause of the trouble finally was traced to a lot of peanuts from Brazil, one of the constituents of the feed, that had been invaded by the fungus *Aspergillus flavus*. Some luck was involved, in that samples of this lot of peanuts still were available, two years or more after the feed in question had been manufactured, and that *A. flavus* still was present in easily detectable numbers in the peanuts and not obscured by a host of other fungi. However, the investigators did establish the fact that a toxin produced by *A. flavus* was responsible for the death of the turkey poults.

Since 1962, nearly 1000 research papers have been published on various aspects of aflatoxins and of poisoning by these toxins. If one estimates the cost of a research paper to be $15,000, which probably is conservative considering the cost of this type of research, this means that at least $15,000,000 has been spent on the investigations of aflatoxins alone during a period of less than ten years. It seems highly probable that other mycotoxins are just as prevalent, and just as potentially dangerous, as aflatoxins, and yet in the same period it is doubtful if as much as 5% of this amount has been devoted to the study of all other mycotoxins combined.

Soon after aflatoxin was discovered in peanut meal in Britain, it was supposed by some that the problem was limited to just a few strains of this one fungus, *A. flavus*, growing in one product, peanuts, in a few areas, such as portions of Brazil and Africa. There was no basis at all for such a naive assumption, but it was comforting to assume that the problem was limited rather than general.

The problem of mycotoxins is not local or sporadic, involving just an occasional chance lot of grains, seeds, or peanut meal or cottonseed meal or fishmeal that happened to become invaded by a rare toxin-producing strain of *A. flavus*, nor is the problem limited to under-developed or developing countries. Rather it seems highly probable that in many of even the highest developed countries of the world, poisoning in animals from consumption of feeds made toxic by fungi is common and widespread. To be sure, the problem or problems may be more prevalent and acute in countries or regions where grains and seeds regularly are stored at moisture contents high enough to be invaded by fungi that produce toxins, and where such fungus-invaded grains and seeds are regularly consumed by people and domestic animals. In portions of Africa, India, and Southeast Asia, severe liver damage is common in the human population, and now is suspected to be caused partly or chiefly by consumption of foods made toxic by fungi. The possibility seems real enough to warrant more attention than has been given to it up to now. The same problems exist even in the most highly developed countries, since all people and their domestic animals, everywhere, are consuming foods invaded by fungi that, under the right circumstances, are capable of producing very potent toxins. The toxins produced by many of these potentially dangerous fungi have not been isolated and identified, and so no means are available for their detection in the foods that we or our domestic animals consume.

The emphasis on increased production of food grains in the over-populated and undernourished countries of the world is, of course, laudable and essential, even though at times it seems to be leading to more people who need more food to produce more people. But increased quantity alone is not sufficient — the people of the developing countries want and need not only more food, but better food, especially a greater variety of nourishing protein foods. Improved *quality* of food is as important as increased quantity. It probably is better, of course, to eat grains, other seeds, poultry, fish, and meats made toxic by fungi than to eat no food at all, but programs to increase food production must include the goal of better quality as well as increased quantity. Since most of the mycotoxins now known are produced by fungi that grow on products *after* harvest, improved storage of food products is essential, especially where invasion of stored products by fungi is an ever-present hazard, and this embraces practically all places where grains, seeds, and feeds are stored. The qualifying "most" and "now known" in the sentence above are included more to emphasize our ignorance than to rule out the possibility that mycotoxins might be formed in grains and vegetables by fungi growing on them before harvest.

A brief summary of some of the major known mycotoxins may help to point up the problem.

THE CHIEF MYCOTOXINS

Aspergillus flavus. Within a relatively short time after the discovery of aflatoxins, A. *flavus* and aflatoxins were found in peanuts in all the major peanut-growing countries, including the United States. Within the last few years aflatoxins have been found in other grains and seeds in the United States, in carload lots of corn destined for use as food or feed, and in other products, such as fishmeal. To anyone who is at all familiar with the fungus, this was only to be expected. Like many other fungi, A. *flavus* probably is about as ubiquitous an organism as can be encountered. It is common in soil, especially in cultivated soil in which crop residues are incorporated, and it is common in all sorts of decaying vegetation, such as grains and seeds whose moisture contents are high enough to permit it to grow — a moisture content in equilibrium with a relative humidity of 85%–90%. In starchy cereal seeds such as wheat, corn, sorghum, and rice this means a moisture content of 18.5%–20%, wet weight basis. And production of aflatoxin is not characteristic of just a few chance and uncommon strains of A. *flavus*, as some have supposed; many strains of the fungus produce aflatoxins, and some may produce other toxins not yet identified. Toxin-producing strains of A.

flavus have been found just about wherever they have been sought — in many kinds of grains and seeds, in peanut meal and soybean meal and fishmeal, in feeds, in black and red pepper, in flour, and in macaroni and spaghetti products. The mere presence of toxin-producing strains of the fungus in all these products does not mean that aflatoxins also are present, since the fungus requires somewhat special conditions to produce appreciable or measurable amounts of toxin — essentially a high enough moisture content and temperature to allow it to grow vigorously for a few days. It is not necessary, however, for material to be thoroughly and heavily overgrown by the fungus for toxin to be present. Brook and White (62) stated that if 3%–5% of a given lot of seeds are invaded by *A. flavus,* the feed made from such seeds may be toxic. If dairy cattle consume feeds that contain aflatoxin, some of the toxin is excreted in the milk, in a somewhat altered form, called aflatoxin M, but still lethal. In Texas Taber and Schroeder (74) grew 213 isolates of *A. flavus* from peanuts, on autoclaved rice and on autoclaved peanuts, and all these isolates produced, on both substrates, some aflatoxin B-1, the most toxic of the several aflatoxins known. This is enough to make a prudent person somewhat dubious about the wholesomeness of almost any product in which *A. flavus* may have grown. As indicated above, it grows in many materials.

How much aflatoxin can be consumed without damage? In animals highly sensitive to it, such as ducklings and rainbow trout, consumption of only a few parts of the toxin per billion parts of weight of the feed, for a few days to a few weeks, may result in severe and sometimes fatal liver damage, or in the development of cancers. Aflatoxin B-1 is one of the most potent cell poisons known, if not the most potent, and is one of the most potent carcinogenic (cancer-causing) agents known. We do not yet know where humans rate in sensitivity, but rhesus monkeys are at least moderately sensitive to damage by aflatoxins. And even though the aflatoxins can be detected and identified, by chemical means, when present in amounts of only a few parts per billion, we do not yet know whether consumption of undetectably small amounts over a period of thirty or forty years might cause damage, including cancer, in those sensitive to them. Our livers may be able to detect smaller amounts than can the chemists. Aflatoxins, by the way, are not inactivated or destroyed by cooking.

As soon as aflatoxins were found in peanuts grown in the United

States, investigations were undertaken by the USDA and by state experiment stations — the work supported in part by the peanut growers
themselves — to determine the extent of the problem and to find out
how to reduce or eliminate aflatoxin in peanuts and peanut products.
It soon was found that the major invasion of peanuts by A. *flavus* occurred after the plants were lifted from the soil and before the nuts
were removed, and in peanuts that were damaged by insects or by
other mechanical breakage. Quick drying after the vines were lifted
greatly reduced infection, and careful inspection of the peanuts before
marketing further reduced the possibility of peanuts infected by A.
flavus getting to the consumer. Thorough testing of peanuts and peanut products, such as peanut butter and peanut flour, for the presence
of aflatoxin just about eliminated the possibility of these materials
being marketed with any detectable amounts of aflatoxin.

A research program also was begun by the USDA to find out whether
any other fungi that might be present in peanuts might produce toxins, but more about this later. Within a few years after aflatoxins were
first recognized as a potential public health problem in the United
States, and before people in general even knew that such a problem
existed, the problem was for all practical purposes solved and the
danger to consumers eliminated. There was no attempt to hide or even
to minimize the potential danger in consumption of peanuts containing
aflatoxin, but fortunately neither were there any scare stories that
might have exaggerated the dangers involved. The whole business was
well and sensibly handled and even when a rather detailed account
of aflatoxins in peanuts appeared on the front page of the *Wall Street
Journal*, it had no effect on the consumption of peanuts and peanut
products in the United States. Perhaps relatively few readers of that
paper carry peanut butter sandwiches for lunch.

In some countries where peanuts are grown and consumed in large
amounts by the population there is no inspection to ensure that heavily damaged or fungus-invaded peanuts are kept off the market, and of
course those grown by people for their own consumption are subject
only to such inspection as the individual happens to give them — and
this probably varies with the available light and the degree of hunger,
among other things. Presumably this would result in the consumption
at times of rather cruddy peanuts, including some invaded to a greater
or lesser degree by A. *flavus* and containing aflatoxin. In Thailand, for

example, where peanuts form an important part of the diet, it is not at all unusual to isolate *A. flavus* from 10%–20% of surface-disinfected shelled peanuts sold for human consumption.

In many tropical countries protein, essential in the diet, especially in the diet of children, is in desperately short supply. Unless protein of sufficient quantity and quality is consumed by children, they may be permanently damaged, both mentally and physically. Various high-protein foods have been developed in Mexico, Guatemala, and some other Latin American countries, and a considerable proportion of some of these foods was at first made up of peanuts, primarily because peanuts constituted a readily available source of protein. Some of the lots of peanuts probably also constituted a readily available source of aflatoxins. To what extent peanuts still are used in such foods is not known to us at present. The possibility also exists that fungi other than *A. flavus* may produce toxins in peanuts, and certainly some research is desirable to ensure that such compounded foods are nutritionally wholesome.

Aspergillus ochraceus. Theron *et al.* in 1966 (75), working in South Africa, reported that several isolates of *A. ochraceus* produced a toxin that caused acute liver injury in ducklings and rats. They isolated and characterized the toxic compound, and found it to be different from, but evidently just as toxic as, the aflatoxins. In work on stored grains and seeds over the past twenty-five years we never have encountered heavy invasion by *A. ochraceus* of a large number of seeds of any given sample from the United States, although some lots of partially spoiled corn in Mexico yielded *A. ochraceus* from 30% or more of surface-disinfected kernels; in some lots of corn there, destined for feed, *A. ochraceus* was a predominant fungus. According to López and Christensen (58), *A. ochraceus* became the predominant fungus in stored corn (where several fungi originally were present) when the grain was kept at a moisture content of about 19.0% and at a temperature of 25° C. for 14 to 32 days; *A. ochraceus* evidently requires rather special conditions to compete successfully with other fungi, and at moisture contents, temperatures, and times other than those given above, it made up only a small proportion of the fungus population. It is not at all unusual, however, to isolate *A. ochraceus* from 3% of surface-disinfected kernels of wheat or corn that have undergone some deterioration in storage. As stated above, according to Brook and White

(62) if a given lot of grain has 3%–5% of the kernels invaded by A. *flavus*, feed made from such grain may be toxic. If the same is true of A. *ochraceus*, ochratoxin may constitute a hazard of unknown proportions in feedstuffs and, in some countries, in foodstuffs — certainly it would be desirable to find out.

Christensen *et al.* (64) in 1967 reported that they found both A. *flavus* and A. *ochraceus* in relatively large numbers (up to 15,000–20,000 colonies per gram of grain) in samples of ground black and red pepper. All the samples of black pepper with which they worked were bought in stores in St. Paul, Minnesota, or were picked up in planes, in restaurants, or in clubs, mostly in the United States. A number of these samples contained more than 100,000 colonies of various fungi per gram of grain; the word "pure" on the labels of containers of black pepper obviously does not refer to freedom from microflora. Some of the isolates of A. *flavus* and A. *ochraceus* from black and red pepper were grown in autoclaved moist corn and fed to rats and ducklings, and resulted in death, with lesions typical of fungus poisoning. Whether these products, which are consumed in only relatively small amounts, but regularly, by those addicted to their use, ever present a health hazard is not now known, but it would be desirable to find out. Methods are now available to detect ochratoxin as well as aflatoxin in materials suspected of harboring them, but so far as we know ochratoxin has not been detected in either foods or feeds. We have bought, by the way, five-pound cans of black pepper from wholesalers, with moisture contents of 19%–20% when received, high enough to permit rather vigorous growth of these and other fungi.

Other species of Aspergillus. In work on mycotoxins at the University of Minnesota, in 1965–67, 196 isolates of *Aspergillus* other than A. *flavus* and A. *ochraceus* were tested for toxicity to rats, chicks, or ducklings. Eighty-five of these, or about 45%, when grown for a time in autoclaved moist corn and fed to the experimental animals, resulted in death in less than seven days (see Plate 10). Some of these species of fungi are relatively common in some lots of some kinds of feed, and their toxin-producing potential should be investigated.

Fusarium tricinctum or F. sporotrichioides. Several species of *Fusarium* have been found to produce toxins of different kinds, some of them extremely potent, and it seems probable that more will be found as work continues. *Fusarium* is a common cause of blight and decay in many kinds of plants, including seeds and fruits. *F. tricinctum* or *F. sporotrichioides* (the names are synonymous) produces a highly lethal toxin, and there is no question about the sensitivity of people to this toxin. Joffe (68) stated: "Alimentary toxic aleukia (ATA) has been recorded in Russia from time to time, probably beginning in the nineteenth century. The disease occurred with special severity in the war and postwar years of 1942 to 1947. In 1944, the peak year, the population in the Orenburg and other districts of Russia suffered enormous casualties . . . more than 10 per cent of the population was affected and many fatalities occurred in nine of the fifty counties of the district."

According to his figures, in eighteen districts of this region, in southwest Russia, there were *1,000 or more casualties per 10,000 of population in that year,* or more than one in ten, due to consumption of overwintered grain that had been made toxic by invasion of *F. tricinctum.* And these were people, not pigs. Yet there still are those who doubt that mycotoxins are or ever can be at all important in the health of man or of his domestic animals. This skepticism is similar to that which prevailed a century ago concerning the possibility that bacteria might cause disease in plants and animals.

F. tricinctum, like its near relative discussed below, requires a period of low temperature to produce its toxin, which is why people were affected in late winter, when they were gathering and eating grain that had overwintered in the field. *F. tricinctum* is not at all uncommon in corn that has been stored on the cob in cribs in the midwestern United States. We have repeatedly isolated cultures of it from such corn, and some of these isolates (sometimes most of them) when grown for a time in autoclaved moist corn and fed to rats were rapidly lethal to the rats.

Fusarium graminearum or F. roseum. The estrogenic syndrome in swine is common in many of the major swine-growing areas of the world. The disease involves various disturbances of the reproductive organs and reproductive processes, and while it is most common in swine it may occur in cattle also. In female swine it results in vulvovaginitis, enlarged uteri, reduced litters, and abortion (see Plates 8 and 9). The cause of this important disease has been unknown, although various agents, including some unknown substance in the feed, have been suspected to be involved. Work at the University of Minnesota over the past few years has shown that a product or products produced by *F. graminearum* (and possibly by other species of *Fusarium*) are one cause, probably the major cause, and perhaps the only cause, of the estrogenic syndrome (65).

Nelson *et al.* stated (71), concerning the estrogenic syndrome in swine:

This involves primarily the genital system of both sexes; in the female, the vulva becomes swollen and edematous . . . and sometimes in severe cases progressing to vaginal prolapse . . . ; the uterus is enlarged, edematous and tortuous, and the ovaries are atrophic, shrunken, and nonfunctioning; pregnant swine may abort. The males undergo a feminizing effect, with atrophy of the testes and enlargement of the mammary glands. This is a fairly common and economically important disease in swine, and until recently the cause was unknown, although the syndrome was reported as early as 1928 to be associated with consumption of moldy corn . . .

Our evidence indicates that a common and perhaps major cause of this disease is a metabolite produced by the fungus *Fusarium*. Corn stored on the cob in cribs commonly is invaded by *Fusarium*, as well as by other fungi. Of the 75 isolates of *Fusarium* from crib-stored corn that we have tested, 35 when fed to rats have resulted in pronounced increase in weight of uteri of the rats in four to seven days. Some of these isolates when grown in autoclaved moist corn and fed to rats, have resulted in death within four to five days, with increases of up to 10 times in weight of uteri. The estrogenic substance produced by *Fusarium* in stored corn has been extracted, purified, and identified. It is a phenolic compound which fluoresces a bright blue when illuminated by ultra-violet light and is identical with one implicated in a similar disease problem with swine in Indiana. Methods of detection and identification of the compound are available. The same compound has been isolated from a commercially prepared feed, that, when consumed by swine, resulted in the estrogenic syndrome, although *Fusarium* no longer could be isolated from the feed. Even though the fungus had died, the toxin remained.

Abortions in swine usually are attributed to leptospirosis, for the diagnosis of which a serological test is available. We commonly encounter a high incidence of abortion in herds that are negative for leptospirosis or brucellosis, or both. These abortions are diagnosed as idiopathic; that is, of unknown cause. In one of our tests, four purebred Yorkshire gilts, all immunized against hog cholera, erysipelas and leptospirosis, and which at the start and finish of the test were negative for brucellosis and leptospirosis, were fed as follows: the controls received normal sow ration; the others received feed containing, respectively, 25, 50, and 100 per cent corn invaded by the fungus. Four days after the test began the gilt receiving a ration containing 50 per cent of corn invaded by *Fusarium* developed symptoms of the estrogenic syndrome, and aborted after 21 days. The control gilt weaned a litter of 10 pigs, and the three gilts fed different amounts of corn invaded by *Fusarium* weaned a total of 11, or an average of 3.7 per gilt.

The estrogenic compound produced by *Fusarium* has been designated F-2. In continued work at the University of Minnesota, a small quantity of purified F-2 was dissolved in liquid and injected into rats every second day for six days; it resulted in greatly enlarged uteri by the end of the sixth day. A smaller amount of the compound, similarly injected, caused a considerable increase in body weight of the animals, but less increase in weight of the uteri. It appears to be more potent when given orally (as it normally would be ingested in feed) than when injected.

So far as our experience goes, to produce the estrogen F-2, corn or other grain (rice is an excellent substrate for this, as it is for the production of aflatoxins) inoculated with the right strain of the fungus must be kept for a time at room temperature, about 20°–25° C., then kept for several weeks at about 8°–10° C. That is, some time is needed at a moderately low temperature for the fungus to produce appreciable amounts of this compound. Some isolates of *Fusarium* that produce F-2, if grown on moist autoclaved corn for two weeks at 20°–25° C., then at 12° C. for several months, are not only strongly estrogenic, but also rapidly lethal to rats and to turkey poults. The effects on larger animals remain to be determined.

It has been stated by some investigators that corn invaded by *Fusarium* sufficiently to be toxic will be refused by swine, and that the swine will starve rather than consume such grain. This may be true of some species of *Fusarium*, but it very definitely has not been true of the estrogen-producing strains with which we have worked — the swine have consumed this toxic corn with relish. Obviously, there are *Fusariums* and *Fusariums*, and so far we are only at the threshold of learning what *Fusariums* may do to domestic animals or man. An isolate of *F. roseum* from muskmelons (in which it causes a rot of the fruit) when grown as described above and fed to rats has consistently killed them in two days; it is one of the most toxic fungi with which we have worked. Do we ever eat muskmelons or cucumbers that have been invaded by this fungus, and if so, are they wholesome?

Fusarium moniliforme. This fungus is common in corn, barley, and wheat, invading the kernels before harvest. It also may produce toxic compounds when grown in the laboratory, but whether it does so in nature is not known. *F. moniliforme* sometimes is present in corn that appears to be perfectly sound — it may grow up through the stem of the growing plant and into the ears and kernels as these form, so that, in at least some varieties of corn, a large percentage of the kernels harbor this fungus at harvest. Probably no great amount of the fungus is present, but how much must be present to make the corn toxic? At present we do not know.

Penicillium. Brook and White (62) list 26 species of *Penicillium* that have been found to produce toxic products when grown in the laboratory. Some of these toxins may be laboratory phenomena. Many fungi when grown on various media in the laboratory will produce materials that when injected into animals will be toxic. In screening tests to detect possibly toxic fungi at the University of Minnesota in 1965–67, 196 isolates of *Penicillium* from various sources were grown in autoclaved moist corn, fed to rats, chicks, or ducklings, and 93, or about 50% of the total number of isolates, resulted in the death in less than seven days of both or several individuals of one or more kinds of animals to which they were fed. Some of these isolates of *Penicillium* resulted in the death of rats in two to three days, after consumption of only a few grams of the fungus-invaded grain.

Penicillium has been implicated as a cause of the hemorrhagic syndrome in poultry, a disease of major economic importance to the raisers of broiler chicks in many parts of the world. In the United States, the hemorrhagic syndrome rates as the third greatest killer of commercially grown poultry, with no known cause, and with no preventive or cure available. About the only positive statement that can be made concerning the hemorrhagic syndrome in broiler chicks is that no known pathogenic bacteria or viruses are involved. Since mycotoxins produced by *Penicillium* have been implicated as a cause of this disease, it may seem strange to an outsider that, with losses of hundreds of thousands of dollars per year involved, no funds and almost no research effort have been devoted to study of this disease. This seems strange to an insider, too.

Grain Storage

Penicillium also has been implicated in a disease of turkey poults, characterized by unthriftiness, loss of coordination, and death; the most characteristic internal symptoms are necrotic lesions in the liver. Usually this disease appears in flocks of young turkeys a week or so old, increases to a peak within the next week or two, then declines and disappears. Usually it does not result in heavy mortality, but at times it may involve 10%–15% of the birds. If 25,000 young birds are started, and 4000 of them die in the first few weeks (as occurred on a turkey farm in Minnesota in 1967), the loss to the grower can be severe.

After the outbreak of this disease on a turkey farm in Minnesota, we were able to reproduce the disease in turkey poults by feeding them on the same ration as was used on the farm. When cultured, this feed yielded many colonies of *Penicillium*. Some of these were grown in autoclaved moist corn for ten days, then dried, ground, and fed to turkey poults a few days old. Depending on the amount of this fungus-invaded grain in their ration, the test turkeys died in five to nine days. The external and internal symptoms were the same as observed in the sick and dying turkey poults on the farm where the trouble appeared. This is strong circumstantial evidence that *Penicillium* is involved in the disease. Much more work will be required to make this possibility a probability.

Japanese workers have good evidence that *P. islandicum*, which is common on rice and which contributes a yellow color to the rice it invades, will cause liver injury and liver tumors in mice. According to Wilson (76), *P. rubrum* produces a potent liver toxin. He stated: "It seems probable that we have merely scratched the surface in identifying the toxigenic fungi likely to be found as food contaminants and the isolation and description of the poisonous metabolites formed by these strains."

Penicillium is abundant in many samples of feed that we have cultured, and often is the predominant fungus in samples of feed. *Penicillium* also is common in many samples of macaroni and spaghetti. Some cheeses, of course, are ripened (or rotted, if you prefer) by species of *Penicillium*. Most certainly, the relation of *Penicillium* to mycotoxicoses, and to the health and well-being in man and his domestic animals, is worthy of some rather intensive and expert investigation. In a report on research on mycotoxins in South Africa Purchase and Theron (72) stated: "Finally, mycotoxin research in South

Africa has indicated the widespread occurrence of significant pathological lesions produced by mycotoxins other than aflatoxin. In preliminary surveys of fungal contaminants of domestic cereal crops, many fungi were found to be toxic, including species of *Penicillium*, *Fusarium*, *Aspergillus* and *Paecilomyces*. One of these fungi, *Fusarium roseum*, is commonly encountered and induces liver lesions which are as severe and extensive as those produced by aflatoxin. This emphasizes the importance of fungi other than A. *flavus*, and toxins other than aflatoxin."

Tables 12–14 summarize data from screening tests for toxic fungi at the University of Minnesota in 1965, 1966, and 1967. Out of nearly 1000 fungus isolates tested, an even 50% were lethal in less than seven days to the rats to which they were fed. The percentage of lethal fungi was higher from corn, feeds, and foods than from peanuts (Table 12). In part this probably resulted from the fact that the samples

Table 12. Number of Fungi Tested at the University of Minnesota and Number Lethal to Rats, 1965–67

Source of Fungi	Number of Isolates Tested	Lethal*	Percentage of Lethal Isolates
Peanuts	396	157	40
Corn, feed, foods	573	331	58
Total or average ..	969	488	50

* Resulted in the death, in less than seven days, of both members of the pair of rats to which it was fed.

Table 13. Toxicity to Rats of Most Numerous Genera of Fungi from Peanuts

Genus	Number of Isolates Tested	Lethal*
Alternaria	23	12
Aspergillus	37	25
Chaetomium	26	11
Fusarium	65	28
Penicillium	80	40
Total	231	116†

* Resulted in the death, in less than seven days, of both members of the pair of rats to which it was fed.
† 50%.

of corn and feed tested were from farms on which severe illness or death of animals was suspected to be due to something in the feed because tests in the Diagnostic Laboratories of the College of Veterinary Medicine had excluded other causes.

Table 14. Toxicity of Most Numerous Genera of Fungi from Corn, Feeds, and Foods

Genus	Test Animal*	Number of Isolates		Percentage of Lethal Isolates
		Tested	Lethal†	
Alternaria	R	60	53	88
Aspergillus	R–C–D	159	60	38
Chaetomium	R	18	15	83
Cladosporium	R	41	19	46
Fusarium	R–T	87	65	75
Penicillium	R–T	116	53	46
Total or average		481	265	55

* C = chick; D = duckling; T = turkey poult; R = rat.
† Resulted in the death, within seven days, of both members of the pairs to which it was fed; or, when more than two animals were used, in the death of all.

Tables 13 and 14 indicate that just about the same genera of fungi from peanuts and from other sources made up by far the major portion of toxic fungi. All these genera of fungi are common in a multitude of materials used for feed and food, and many isolates of them have the ability to produce metabolites toxic to laboratory animals when ingested. There are a number of limitations attached to screening tests such as these (as there are to almost any tests of toxicity of these fungi). But the evidence from these tests, as well as from similar or comparable tests made wherever mycotoxins have been worked on at all thoroughly, indicates that the problem of mycotoxins in human health and in the health of our domestic animals is of greater magnitude than is generally realized even by those most concerned with maintaining health in people and in domestic animals throughout the world.

GENERAL ASPECTS OF RESEARCH ON MYCOTOXINS

If mycotoxins are as important as is suggested by some of the evidence presented above, why have they not been investigated more? No simple answer can be given.

Really effective work on fungus toxins and on the diseases caused by them can hardly be undertaken by an individual — and research men are notoriously individualistic, each choosing to follow his own star, or his own will-o'-the-wisp. Teamwork is required. A mycologist must be involved who knows his fungi, how to isolate them and how to grow them. This can be done routinely, of course, and then will give routine results, but if done imaginatively it may give results that are not at all routine.

A given sample of feed suspected of harboring toxin-producing fungi and fungus toxins, may, when cultured on agar media, yield large numbers of fungi. Any one agar medium is inevitably selective to a certain extent, and so a number of agar media must be used. From a given sample of feed, a dozen or more genera of fungi may be isolated. One isolate or one strain of a given fungus may produce a toxin, another not, and so it is necessary to isolate several to many cultures of each of the fungi. Whether a fungus capable of producing a toxin will or will not produce the toxin depends upon a number of things — the substrate in which it grows, the temperature of incubation, the time of incubation, the amount of aeration, and so on. So from a given sample of feed ten isolates of each of ten genera may be selected. Each of these must be grown on a suitable substrate for a time, then fed or otherwise administered to experimental animals. Grown on what substrate and for how long, and administered how to what animals?

In our own work we have used a substrate of autoclaved moist corn, or of corn and rice, mostly because corn makes up a major portion of many kinds of animal feeds throughout the world and rice is known to be a good substrate for production of toxins by some fungi. Usually we have incubated the inoculated material for about ten days at room temperature, followed by two to three weeks at 8°–10° C., because we have found that such a schedule will result in toxin production by many isolates of many kinds of fungi. In Minnesota and in regions of similar or comparable climate, this seems to be a realistic schedule, because not uncommonly fungi grow in corn for a time at moderate temperatures in the fall, then at lower temperatures during the winter. Also diseases that appear to be mycotoxicoses are more prevalent in midwinter and late winter than at other times of the year. Such an incubation schedule may not be realistic for some fungi and other climates.

After incubation the now heavily fungus-invaded material must be administered to animals, and the effects determined. This phase of the work must be performed by or under the direction of a competent veterinarian, one who is thoroughly familiar with the animals, and who is competent to evaluate such lesions and symptoms as are evident in both premortem and postmortem examination. Since we are interested mainly in toxins that produce their effects when consumed as feed, we prefer to feed the test materials to test animals, at least in the preliminary screening tests. Each of the samples is fed, either as the sole ration or as a mixture with normal feed, to rats, turkey poults, ducklings, chicks, or, sometimes, swine. Ordinarily we aim to use, as a test animal, that kind of animal in which the disease has appeared on farms, but sometimes this is not possible, or at least is difficult with the facilities available. If we are interested in a disease of dairy or beef cattle which we suspect to be due to fungus toxins, we should use cattle as the test animal. This means growing enough of the fungus to form at least a fair portion of the diet for several cows for at least some weeks; it also means purchasing a dozen or so cattle (since some must serve as controls), and paying for their housing and care. This runs into money, far more than we have had available, and far more than we are likely to have available in the foreseeable future. So in that case we use rats as the test animal, even though we recognize full well that rats are not grown as a food animal, and that what is toxic to a rat may not be toxic to a cow, and that what is not toxic to a rat may very well be toxic to a cow. This does not make for complacency or self-satisfaction in the work.

Some workers with mycotoxins grow the fungi with which they are concerned on liquid media of one sort or another, then inject some of the liquid or a purified fraction of it orally or parenterally into the animal, or inject it into fertilized eggs (death of the embryo indicates the presence of toxin), or apply it to the shaved skin of a mouse or rabbit (inflammation, a welt, or death of the animal indicates the presence of a toxin). It has seemed to us that in work designed to find out whether mycotoxins in feeds are important in the health of domestic animals, it is preferable to grow the fungi on constituents of common feeds and to administer them by feeding, as they would normally be consumed.

Once a given isolate of a given fungus has been found to be toxic to a given kind of test animal, or preferably to several kinds of test animals, in repeated tests, the work has only begun. It is desirable, and indeed necessary, if any real progress is to be made in the solution of the problem, to isolate, purify, and establish the nature and characteristics of the toxin or toxins. For this a physiologist or biochemist and also an organic chemist are necessary. This work requires expensive equipment and a good deal of expertness, but only by actual extraction and identification of a given toxin of known characteristics can we tell whether a given toxin is present in a suspect lot of feed or food. There is no other way. The mere presence or absence in a suspect lot of food or feed of a fungus known to produce a toxin is of no value whatever in judging whether toxin might be present — the toxin must be extracted and identified.

Techniques now are available for the extraction, purification, and identification of several kinds of aflatoxin, produced by *A. flavus*; for ochratoxin, produced by *A. ochraceus*; for the F-2 toxin, produced by *Fusarium*; and for rubrotoxin, produced by *P. rubrum*. Many expert investigators contributed to the development of these methods, and the methods are still being improved and refined. The toxins produced by some fungi appear to be exceedingly difficult to extract and purify, but until these toxins are purified and identified, and techniques worked out for detecting their presence in feeds and foods, we are in the dark, because we have no way of detecting their presence.

The discussion above should indicate some of the difficulties and frustrations inherent in the work with mycotoxins and mycotoxicoses. Probably many decades of work lie ahead, by teams of able and dedicated investigators, before we will have even the major answers we need to evaluate the role of even the more common mycotoxins in the health of our domestic animals throughout the world, and in the health of mankind.

7 EVALUATION OF CONDITION AND STORABILITY

A practical means of evaluating the condition and storability of given lots of grains and seeds would be of great benefit to all those concerned with marketing, storage, and processing of grains and seeds and their products. So long as the causes of deterioration were unknown, and grain was thought to have an "urge to heat and germinate in the spring" regardless of where or how it was stored, or some lots of grain were thought to have mysterious but unexplained "inherent tendencies" to spoil, little progress could be made in predicting storability. Until very recently grain elevator personnel did not even have a means of determining accurately the moisture content of the grain they bought and stored (the moisture meters were not accurate), much less a means of evaluating the condition of grain as they bought it or as it was stored. Some elevator men had much better records than others in preserving the quality of the grains they stored, but the reasons for this were not known. And so long as increases in respiration, in fat acidity value, and in temperature of stored grains and seeds were thought to be due solely to physiological activities of the grains and seeds themselves, there seemed to be no rational approach to prevention of spoilage.

Most of this book up to this point has been devoted to presentation of evidence that, in the absence of insects, and sometimes when they are present, fungi are a major cause of spoilage in stored grains. This has now been established beyond any doubt. If, therefore, increases in storage fungi could be detected before they develop enough to cause appreciable damage, it should be possible to prevent most of the damage. Or, since we now know fairly precisely what conditions are necessary for the growth of these fungi, and what conditions will

94

greatly retard their growth, it should be possible to maintain conditions in any given bulk of grain unfavorable to the development of storage fungi, and so prevent spoilage. According to circumstances either or both approaches may be used.

INDIRECT METHODS OF MEASURING FUNGUS INVASION

Fat acidity value (FAV). In Chapter 5 evidence was presented to indicate that increases in FAV in stored grains are likely to be due chiefly to storage fungi, not to activities of the seeds themselves. If increase in storage fungi is regularly accompanied by increase in FAV, measurement of FAV in a sample of grain should be a quick and easy way to evaluate increase in storage fungi and to judge relative storability. Considerable work has been done on this, the results of which are summarized in the following section.

Black and Alsberg (2) in 1910 and Besley and Baston (78) in 1914 suggested that acidity of corn be used as a measure of soundness and quality of the grain. They were concerned with total acidity of corn. Zeleny and Coleman (85) in 1938 proposed that the amount of free fatty acids present in a given sample might be a useful criterion of the condition of the lot from which the sample was taken. In 252 samples, they found that FAV increased as grade decreased. Some of their data are summarized in Table 15.

In a subsequent paper (86) they published additional evidence concerning the relation of FAV to grade in corn. Some of the data from this paper are summarized in Table 16.

FAV never has been accepted as a grading factor in corn or in other grains, for one reason because of the great range in FAV within any one grade. Some samples of grade No. 5 have a lower FAV than some samples of grade No. 1; so if FAV were to be the major factor in determining grade, the other characteristics that have long been used would have to be disregarded or would have much less weight. Some other qualifications have limited the value of FAV as a major grading factor in corn: (1) In a given year many truckloads of corn come to market that by all the standards now used to evaluate grade and quality are good No. 2 grade, and yet they have FAV higher than the limit proposed for No. 2 corn. (2) Some of the fungi that invade wheat and corn stored with moisture contents below 15.0%, and that cause loss in germinability, increase in discolored germs, and mustiness, do not cause any detectable increase in FAV. In our experience much of the deterioration that occurs in stored wheat and corn occurs in the lower range of moisture content where storage fungi can grow — from 14.0% to 15.0% or 15.5% in the starchy cereal grains, that is, in the range in which there may be little or no increase in FAV with increasing invasion by the fungi. (3) Cracked and broken kernels of

Table 15. Grade and FAV of 252 Samples of Corn

Grade	Damaged Kernels (in Percentage)	No. of Samples	FAV (Average)
1	0–3.0	88	17
2	3.1–5.0	46	26
3	5.1–7.0	28	36
4	7.1–10.0	31	46
5	10.1–15.0	28	53
Sample	15.0+	31	125

Source: Zeleny and Coleman (85).

Table 16. Range in FAV and in Germination Percentage in Corn of Different Grades

Grade	No. of Samples	Range in FAV	Range in Germination Percentage
1	21	14–39	42–90
2	43	13–50	2–93
3	28	22–63	16–86
4	31	21–70	1–85
5	28	27–91	0–77

Source: Zeleny and Coleman (86).

corn may increase greatly in FAV without being invaded by fungi, and while this may reduce quality of the grain for some uses, it does not do so for others.

High FAV unquestionably indicates some deterioration, either biological or purely chemical, but deterioration may occur without any significant increase in FAV. Other things being equal, grain of low FAV is preferable to that of high FAV, but FAV is of limited value as an indicator of over-all quality of grain.

Respiration. Measurement of respiration of stored grains has been suggested as a means of determining what is happening beneath the surface of large or small bulks, and theoretically it should be a good indicator of biological activity. However, if carbon dioxide production of a total mass is measured, one cannot tell whether slow respiration is going on uniformly throughout the mass, or whether in certain small or not-so-small portions of it insects or fungi are respiring rapidly. So far as spoilage is concerned, this makes a great difference. Also, fungi may be growing so slowly that there is no detectable increase in respiration. No one, so far as we are aware, has ever measured the rate of respiration of members of the *Aspergillus restrictus* group growing in wheat or corn with a moisture content of 14.0%–14.5%, the level at which these fungi are likely to predominate. Any method of evaluating fungus activity in grain that failed to detect the slow and insidious increase of *A. restrictus* would be of limited value, and might give a false sense of security. Regardless of its possible theoretical value, we do not know of any case of incipient spoilage having been detected in commercially stored grains by means of measurement of respiration rate.

Temperature. Moderately dry grain is a good insulator; according to Oxley (82) about a third as good as cork and ten times as good as concrete. This low thermal conductivity of grains means that if heat is generated within a bulk of grain, it will be dissipated very slowly, and so may accumulate, with the eventual formation of hot spots. If insects or fungi are developing at or near their optimum rate in grain, each kind of insect or fungus will raise the temperature to about the maximum that it can endure — in the case of insects, about 40° C. (104° F.), and in the case of fungi, up to 55° C. (131° F.).

An increase of temperature within a bulk of stored grain means that either insects or fungi, or both, are developing. Temperature in stored grains usually is measured by means of thermocouples attached to cables installed permanently in bins, the thermocouples spaced about 6 feet apart vertically on each cable, and the cables separated from one another by 20–25 feet. This means that there is one thermocouple per 2000–3500 cubic feet of grain. If insects or fungi begin to develop in a volume of 20 bushels equidistant from the nearest thermocouples, they may raise the temperature as high as the insects or fungi can endure without this being indicated on the thermocouples until the spoilage spreads so that the temperature rises in the grain near the thermocouples. That is, a detectable rise in temperature within a bulk of stored grain is likely to indicate, not incipient spoilage, but advanced spoilage somewhere near where the rise in temperature is detected. In stored malting barley, any detectable rise in temperature is regarded as an emergency to be taken care of at once, while with some other grains and some other grain handlers the temperature is allowed to increase greatly before the man in charge becomes sufficiently alarmed to do anything about it. In the one case, loss is avoided by taking care of the trouble before it becomes acute, and, in the other, loss is suffered because the warning signals are disregarded. Temperature detection devices attached to a pneumatic probe make it possible to measure temperature any place in a bin of grain that the operator wishes to.

Temperature detection devices are very useful aids in maintaining quality of grains in storage, but they are only an aid, and must be so recognized. They can only indicate that, if the temperature rises, spoilage is in progress. Fungi such as *Aspergillus restrictus* and A. *glaucus* growing in corn, wheat, or rice with a moisture content of 14.0%– 14.5% will not respire rapidly enough to cause any detectable rise in temperature, but they may cause total loss in germinability and a great increase in germ damage. They may slowly increase the moisture content of the portion of the grain in which they are growing, until it reaches a level where A. *candidus* and A. *flavus* can take over, in which case rapid spoilage will be in progress in a few days. This may be indicated by a slight rise in temperature at a thermocouple several feet away, and the elevator manager, if he happens to have a high threshold of alarm, may decide to do nothing about it just then, since the rise in temperature is not great enough to alarm him. If he were to take samples from the grain in and around the heating portion, he would know the size of the hot spot and the condition of the grain around it.

Germinability. Decrease in germinability is one of the most sensitive indicators of incipient spoilage — too sensitive, in fact, since, as indicated earlier, germinability may decrease greatly without any accompanying decrease in processing quality. However, in parcels of grain stored for months or years, decreasing germinability with increasing time may be an indicator that something is wrong — most probably that the grains are being invaded slowly by storage fungi, and that these might be increasing gradually to the point where their further increase can be sudden, rapid, and disastrous. Germination tests are not difficult to make, and at times they give useful information, especially when used in conjunction with other tests.

DIRECT MEASUREMENT OF FUNGUS INVASION

Microscopic examination. One cannot detect the beginning invasion of grains by storage fungi simply by examining the grains microscopically, but early invasion can be so detected in many grains before the fungi cause sufficient damage to result in any reduction in grade. When the invasion becomes microscopically visible, the embryo is weakened or dead, but is not likely to be discolored. The discoloration develops later, when invasion has progressed to the point of decay. If perithecia develop on the surface of the embryo, it means that the grain has been stored at a moisture content of 15.5% or higher for at least several weeks, because only at those moisture contents can the fungus or fungi involved form these characteristic structures.

This has been established beyond question. We have received numerous samples of wheat and corn in which severe deterioration had occurred, but which those in charge of the grains claimed had been stored at moisture contents below 13.0%, maintaining therefore that the spoilage was due to mysterious causes; perithecia were abundant on the surface of the germs of these samples, positive proof that they had been stored at moisture contents of 15.5%–17.5%. What the men in charge of these samples, or of the bulks from which the samples came, meant when they said the grains had been stored throughout their storage life at moisture contents below 13.0% was that the moisture contents as shown on the warehouse records were below 13.0%. The moisture content of the grain in the bins was unknown to them, because they never had taken samples of it to determine what was going on.

Fungi develop on the outside of seeds too, but are likely to be scoured off when grain is shifted or transferred. In seeds without internal cavities, such as beans, peas, and soybeans, the early stages of invasion cannot be readily detected by microscopic examination, but can be detected by culturing the seeds on an agar medium, as described below.

Culturing. The usual way to detect the presence of fungi in seeds as well as in most other materials is to place them on an agar medium on which the fungi will grow out and so reveal their presence. (See Plate 1.) Almost any agar medium is selective to a certain extent, in that it will allow some fungi to grow, but not others, or will allow some fungi to grow much more rapidly than others and so conceal the slow-growing ones. Since storage fungi grow at moisture contents in equilibrium with relative humidities of about 70% to 90%, they are growing without free water, and in an environment of rather high osmotic pressure. Some of them, especially those in the *Aspergillus restrictus* and *A. glaucus* groups, not only endure, but *require* a high osmotic pressure to grow well. Therefore most media designed to culture these and other common storage fungi include salt, such as sodium chloride, in concentrations of 5% to 20%, or sugar, such as sucrose, in concentrations of 20% to 65%.

To determine the numbers and kinds of fungi present is relatively simple: 50 or 100 seeds or kernels are shaken for one minute in 2% sodium hypochlorite, a common household bleach and sanitizing agent available in almost any grocery store in any country where groceries are sold in stores. They are then rinsed with sterile water and, with sterile tweezers, placed on an agar medium in petri dishes, 20 to 50 seeds per dish. The dishes are incubated at about 27° C. for five to seven days, then examined with the aid of a dissecting microscope and a good light, by someone who knows the different fungi.

Plate 7 illustrates the difference in number and kinds of fungi isolated from two lots of sorghum seed of different germinability. In this case, the storage fungi have damaged only the germinability; the interior of the seeds was not discolored, and none of the seeds were decayed. Outwardly, most of the seeds still appeared to be sound, although sporophores of *Aspergillus glaucus* were growing from the point of attachment or from the embryo of some of the seeds. These seeds had been removed rather carefully from the bins where they had been stored, and so the delicate external sporophores of *Aspergillus* had not been scoured off. Had the grain been emptied from the bin in the regular fashion, there would have been no outward indication, even microscopically, of the presence of the fungi. The culturing technique, however, would still have revealed the presence of the fungi.

Note that there has been a complete change in fungi in these samples, from nearly 100% of the kernels yielding field fungi and 0% yielding storage fungi in those of high germination, to 0% yielding field fungi and nearly 100% yielding storage fungi in those of low germination. This drastic and obvious change in numbers and kinds of fungi occurred *before* there was any detectable loss in quality other than decrease in germinability. Had the seeds been tested earlier, decrease in field fungi and increase in storage fungi would have been found to precede decrease in germinability also.

The idea of using number and kinds of fungi isolated from a sample of grain, in conjunction with other characteristics discussed above, to evaluate condition and storability of the lot from which the sample was taken is not new or untried. In 1952, Del Prado and Christensen (80), working with rice, stated: "If a given seed lot has a very high count of a species of fungus known to invade seed and to cause loss of viability in storage, this is presumptive evidence that the seed is deteriorating or has already deteriorated. If a fungus known to grow at the moisture contents normally encountered in storage is found in a large proportion of the seed, and if the total count is high, and the viability of the seed is below normal, it may be presumed that this lot of seed is a poor storage risk. At present there is no microbiological evaluation of any agricultural seed when it goes into storage. If practical methods of microbiological evaluation of the quality and storability of seeds could be developed, they might be of value for many kinds of seeds." Such methods have been in use in the Grain Research Laboratory of Cargill, Inc., since 1952.

STORABILITY RATINGS

Sorger-Domenigg *et al.* (83), working with wheat, in 1955 stated: "A combination of tests, involving moisture content, number and kinds of molds present, viability, and fat acidity should serve to predict storage behavior and extent of actual damage. The moisture content should indicate whether there is present or future danger, mold tests whether invasion of the seed already has occurred, viability whether incipient deterioration has developed, and fat acidity should give some measure of the actual damage which already has occurred."

By this time, at least one large grain merchandising firm and one malting firm were already using tests of number and kinds of fungi in samples drawn from given lots of grain as a means of evaluating condition and storability of the lots from which the samples were taken. The use of such tests has spread slowly, partly because few grain storage firms have research or control laboratories where such work can be done, but mostly, perhaps, because of the failure of men in the upper echelons of management to realize the value of such tests in the maintenance of quality.

Qasem and Christensen (22) compared the relative rate of deterioration of commercial lots of grade No. 2 corn with that of high-quality, hand-shelled corn, stored at 16% moisture content and 25°C. for two and four months. The idea of this was that the major differences in storability or rate of deterioration between the No. 2 lots and the high-grade corn might be in such measurable characteristics as germinability, mechanical injuries, degree of invasion by storage fungi, amount of broken kernels and debris present, and, of course, the moisture content at which each had been kept since harvest. Different lots of No. 2 corn differ in these characteristics also. If these characteristics determine condition and storability, it should be possible to see these differences most clearly in a comparison where the differences were relatively large, as between the hand-shelled, high-grade corn and the lots of No. 2 grade.

Some of their results are summarized in Table 17. They said, "After 2 months the germination percentage of the 3 commercial samples ranged from 1 to 12%, and the percentage of discolored germs (brown and ochre together) ranged from 16 to 46%. The seed quality corn germinated 98%, 2% of the germs were ochre colored and none of the germs was brown. After 4 months the germination of the commercial samples was zero, and discolored germs ranged from 42 to 86%; the seed quality corn germinated 69%, 4% of the germs were brown and 14% ochre. Under conditions that permit storage fungi to grow, corn as ordinarily encountered in commercial storage often can be expected to deteriorate much more rapidly than the high quality seed corn used in most of our tests."

They injured the pericarps of corn kernels that were otherwise of seed grade, then exposed these injured seeds, along with noninjured seeds, to conditions favorable for the growth of storage fungi. The injuries consisted only of scratches that penetrated into or through the rather thick pericarps of corn seeds, and were produced by shaking the kernels in a two-pound coffee can whose bottom had been perforated in many places by pounding a nail through from the outside in, leaving many small but sharp projections. The idea behind this was that Koehler (81) had examined many lots of seed grade corn over a period of ten years and found that an average of 78% had pericarp injuries detectable by examination after the corn had been stained with a 0.1% aqueous solution of fast green. Koehler was interested in the possibility that such injuries might facilitate invasion of the seed by fungi when the seed was planted, and so make for more seedling blight. Qasem and Christensen thought that such injuries might provide invasion sites for storage fungi. They stored the samples for two months at 85% relative humidity and 25° C., then examined them. They said: "The differences between the scarified and non-scarified portions of each lot were large: the germination percentage of the scarified grain was much lower than that of the nonscarified, percentage of discolored germs much higher, percentage of surface-disinfected kernels yielding *A. candidus* was from 2–3 times as great, and the mold count 2–3 times as great. We conclude from this that

Table 17. Germination, Germ Discoloration (Brown), and Fungus Invasion of Three Commercial Samples of Grade No. 2 Yellow Dent Corn and One Sample of Seed-Grade Yellow Dent Corn Stored Two and Four Months at 16% Moisture Content and 25° C.

Sample and Months Stored	Germination Percentage	Percentage of Brown Germs	Percentage of Surface-Disinfected Kernels Yielding *Aspergillus Glaucus*
Commercial sample 1			
0 months	50	1	40
2 months	12	2	80
4 months	0	16	100
Commercial sample 2			
0 months	62	2	55
2 months	1	6	100
4 months	0	18	100
Commercial sample 3			
0 months	48	5	46
2 months	6	10	100
4 months	0	28	100
Seed grade			
0 months	98	0	4
2 months	98	0	48
4 months	69	4	100

Source: Qasem and Christensen (22).

cracks, breaks, or scratches in the pericarp over the germ furnish a ready port of entry for storage fungi."

Probably most lots of No. 2 corn encountered in commerce would have cracks or scratches through the pericarp in a large percentage of the seeds, although this may not be inevitable. The main point of the comments above is that such scratches, breaks, or cracks constitute another factor in "condition" of corn, and also that decrease in germinability and increase in discolored germs are a product of invasion by fungi, which is a measurable characteristic. The authors emphasized:

The rate at which corn deteriorates in storage is influenced greatly by a number of factors, of which the most important appear to be moisture content, temperature, the kinds of fungi involved, the length of time and the rate at which the fungi grow, the condition of the grain, and the number and perhaps the severity of injuries in the pericarp of the corn kernels. These factors operate together, at least in corn as it ordinarily is encountered in commercial storage, and so must be considered together. This probably explains why some lots of corn can be stored at a moisture content of 13–14% for several years without an appreciable amount of germ damage developing, while in others, stored at essentially the same moisture content and temperature, extensive germ damage occurs; or why some corn bulks stored at 15% moisture content and 50 F retain for a long time the same condition as when they were binned, while others after a short time undergo molding and heating.

They also stressed:

The major cause of germ damage in corn in ordinary commercial storage is the invasion of germs by storage fungi. The dark brown germs typical of so many cases of germ damage encountered in commercial lots of corn appear to be produced mainly by a combination of invasion by any one of a number of storage fungi plus a comparatively high temperature. A knowledge on the part of grain men of the degree of invasion of given lots of corn by these storage fungi, as well as more accurate knowledge of and better control of both moisture content and temperature in bulks of stored corn would aid greatly in reducing or avoiding losses caused by storage fungi.

The evidence was accumulating to the effect that fungi were a major factor in the spoilage of stored corn and other grains, that the invasion of grain by these fungi was influenced or determined by such measurable things as temperature and moisture content and time, and that if the grain in storage could be kept at a temperature and moisture content that would not permit storage fungi to invade it to any great extent it could be kept for years without damage. One obvious way to determine whether storage fungi are increasing in a given bulk or parcel of grain is to take samples from different portions of the grain at intervals and test them for numbers and kinds of fungi. If moisture content and temperature of the samples are determined at the same time, and the grain examined for the presence of fungi and for damage, this gives information about the conditions that have prevailed in different parts of the bulk, and information about whether damage might be developing here and there that otherwise might go undetected.

Following this approach, Welty *et al.* (84) took samples at depths of three, nine, and twelve feet at each of seven locations in a bin of 100,000 bushels of corn, at intervals of three to six months, for four years. At some of the test periods samples were taken from a depth of six feet also. The bin was a quonset type, and the total depth of the grain was just over twelve feet. The bin was equipped with fans that permitted the operator to cool the grain to 40°–50° F. shortly after the bin was loaded and, by running the fans at appropriate times, to maintain essentially this temperature throughout the mass throughout the storage life of the grain.

The corn that went into the bin was of grade No. 2, but was in uniformly excellent condition, since the first 21 samples taken, some months after the bin was loaded, averaged 93% germination, none had a moisture content over 14.5%, none of the surface-disinfected kernels yielded storage fungi, and no discolored germs were detected. No. 2 corn of this quality may be somewhat rare, but at least this operator had 300,000 bushels of it that year, all of it equally good.

The corn was aerated twice a year, at a time when the temperature and humidity were moderately low, and, as stated above, temperatures of 40°–50° F. (5°–9° C.) were maintained throughout most of the grain (except surface layers) throughout its storage life. The condition of the grain at a depth of six feet in the bin after storage for 33, 39, and 52 months is shown in Table 18. There was a slow invasion by storage fungi of the grain at a depth of three feet, accompanied by a reduction in germination percentage and by development of some slightly discolored germs, although no damaged germs or kernels were detected.

When the grain was loaded out and sold after storage for more than four years, it still was above the average of grade No. 2 corn.

The taking of samples from this bin required a few hours every three to six months, and the testing of these samples at each test period required at most only a portion of about three man-days spread over a week. The results of such sampling and testing enabled the man in charge of the grain to *know at all times* the condition of the grain in all portions of the bin, and to continue this grain in storage from one test period to the next with absolute assurance that its good condition would be maintained. It was, in fact, on the basis of these tests that the grain was continued in storage as long as it was. And it was also on the basis of these tests that the decision was made to sell the grain at a period of high price in the spring, rather than hold it for another year and risk the possibility of some damage developing in the upper three feet of the grain.

Table 18. Characteristics of Corn Taken from a Depth of Six Feet in a Commercial Bin after Storage for 33 to 52 Months (Each Figure Being an Average of Seven Samples)

Characteristics	33 Months	39 Months	52 Months
Moisture content (in percentage)			
Average	13.8	13.6	13.5
Range	13.4–14.5	13.4–13.9	13.1–13.8
Temperature (in degrees F.)			
Average	50°	41°	43°
Range	47°–54°	40°–43°	
Germination percentage			
Average	81	78	90
Range	52–91	48–90	80–97
Ocher germs (in percentage)	0	2	2
Surface-disinfected kernels yielding storage fungi (in percentage)			
Average	10	14	22
Range	0–28	0–62	8–78

Source: Welty *et al.* (84).

The authors stated:

There were no unexpected or sudden changes in condition of the grain, no "mysterious" development of germ damage such as some grain men still believe may occur in corn stored at moisture contents of 12.5–13.0%, and certainly none of this corn ever displayed the slightest "urge to heat and germinate in the spring."

The periodic sampling gave a reliable measure of the condition of the corn throughout the bin at the time when samples were taken, and also permitted an accurate evaluation of the storability of the corn up to the next sampling period. All the evidence indicates that if corn is sound and free of storage fungi when stored, of uniform moisture content no higher than 14.0–14.5%, and is cooled and kept at a uniform temperature of about 45° F, it will remain in excellent condition for at least 4 years, and probably longer.

This has been borne out by the experience of Cargill, Inc., which has repeatedly stored in its Chicago terminals bulks of up to 3,000,000 bushels of corn for as long as three years with no decrease in grade. But its officials have known at all times the condition of the grain in all parts of the bulks, simply by periodically withdrawing samples and testing them for the characteristics that reveal storability. Deterioration and spoilage in stored grains usually develop when the condition of the grain is not what the man in charge of it assumes it to be. Sampling and testing enable him to know the condition of grain throughout his warehouse. There is nothing mysterious or esoteric about it, simply sampling and testing that any moderately equipped laboratory can do, plus common sense.

8 DRYING, AERATION, AND REFRIGERATION

Much grain is harvested with a moisture content too high for safe storage. Sometimes this is a result of moist weather at harvest, but in regions where the relative humidity is high at night, with or without the deposition of dew, even in relatively dry weather grain harvested in early morning may have a moisture content 5%–10% above that harvested in midafternoon. We found differences greater than that in sorghum seed collected from different parts of the same heads at the same time — the grain collected about 8:00 A.M. from the top of the heads of several plants had an average moisture content of 16.3%, whereas that from the bottom of the same heads had an average moisture content of 35.0%, a difference of almost 20%. Few people seem to be aware of this source of variation in moisture content, although at times it can have a great influence on the storability of a given lot of grain.

As noted in Chapter 1 most of the corn marketed in the United States is harvested with picker-shellers or combines, and for best results the picker-sheller requires a moisture content in the corn of about 23%. Unfavorable weather at harvest time, delayed maturity, or other factors may result in corn being harvested with a moisture content of 25%–30% or more. In much of the corn belt of the United States the daytime temperatures during harvest are high enough to promote rapid growth of fungi, and if the moisture content of the corn is above about 22%, yeasts and bacteria may grow rapidly too. This makes for grain of high storage hazard.

Several approaches and combinations of approaches have been developed to improve storability and maintain marketable quality in corn and other grains harvested with a high moisture content. Chiefly there are three: (1) drying to a moisture content safe for storage; (2) aeration, to maintain a low and uniform temperature so as to prevent later migration of moisture; (3) refrigeration, which is only aeration with artificially cooled air. Each of these seems simple, and in principle all of them are so, but in practice they are far from simple. The operator must take into account such things as the moisture content of the grain when he receives it, its condition, how long it can be kept before drying without losing grade and quality, at what temperature and to what moisture content it should be dried for a given length of storage, the uses to which the grain is to be put, shifts in market prices and demands, and above all, costs. To achieve a balance among all these sometimes conflicting and shifting factors is not easy, and calls for expert judgment, as will be evident.

DRYING

Probably ever since grains have been grown, that portion of the harvest intended to be used for planting of the next crop has been given special care; certainly in many regions of the world where grains constitute a major food such care would be essential if the grain was to retain its desired germinability.

This special care included drying, where necessary, at first (and in some places still) in the sun out-of-doors, and more recently by artificial heat. Papers dealing with artificial drying of corn were published in the 1920's, and even before that there was some investigation of the effects of temperature on the germinability of grains and seeds. From the 1940's on, much research was devoted to the problem of grain drying, and the chapter by Hukill (88) on drying of grain in the book *Storage of Cereal Grains and Their Products*, published in 1954, cites 77 references.

Drying, by whatever means it is done, is only an aid to maintenance of quality in stored grains and seeds, and is by no means a cure-all for storage problems. Grains are hygroscopic, and once dried they can later gain moisture from seepage, leakage, and moisture shifts, from production of moisture by insects, mites, and fungi, and from contact with air of high humidity. The drying process itself involves many complications, not the least of which is that of cost.

The moisture in grain is worth the same price as the grain itself, since grain is bought and sold on a wet weight basis. When corn sells for $1.00 a bushel, or about 2¢ per pound, the water is, of course, worth

2¢ per pound also. If soybeans sell for $3.00 per bushel, the water in the soybeans is worth 5¢ per pound, or about 10¢ per quart, or 40¢ per gallon. This is more than double the cost of production of a gallon of gasoline, and about the same as the cost of production of an equal amount of industrial alcohol or of whisky. It is for this reason that when corn is sold with a moisture content over 15.5%, the upper limit for No. 2 corn in the United States, the seller receives 2¢–3¢ a bushel less for each 1.0% of moisture above 15.5%. However, no premium is offered for corn with a moisture content *below* 15.5% and so if a batch of corn is dried to a lower moisture content, the owner of the corn will lose 2¢–3¢ for each 1.0% of water driven off below a moisture content of 15.5%. Also the fuel required to remove the water, and the dryer and other machinery required to handle the grain before, during, and after drying, as well as the manpower to run them, are expensive. Bailey (1) stated:

There are several dangers in operating grain driers, but the principal one is that it will be operated at a loss. This occurs because practically everyone either overlooks or underestimates items of cost. These need to be carefully considered.

The three main items of cost in drying are shrinkage, fixed cost and operating cost. Errors in cost of shrinkage come from any or all of these sources:

1. Inbound moisture content. If we pay for 20% moisture corn but actually receive 23%, we have already sustained a severe shortage besides the extra condition risk and drying time involved. Most moisture testers are not accurate above 22%. They should be checked regularly.

2. Underestimated shrink. Shrink always exceeds the moisture change, and dry matter is blown out during drying. The only certain measurement of shrink is to weigh the grain before and after drying, but that in itself is often difficult and expensive. It can be estimated from the moisture content before and after, if it is accurately known. There are tables and slide rules for this purpose, and if available, should be used.

3. Underestimating cost of shrink. This cost is the amount of shrink times the value of the grain *after* drying, not before. Or, it is the dollars paid for the grain divided by the amount left *after* drying, not before. It should be remembered that the cost of the shrink depends on what we paid for it, so that as the price of corn goes up, standard discounts become inadequate.

Drying costs are also subject to error and omission. They consist of the following items:

1. Fixed costs. The fixed cost per bushel is the annual cost of the drier and all related equipment including handling facilities, and con-

sists of depreciation, interest, insurance, taxes, etc., divided by the annual total of bushels *after* drying not before.

2. Operating costs. These are costs of heat, labor, power, repairs, supervision and inspection applied to bushels after drying, not before.

3. Storage space. Whether or not it is needed for something else, it is a part of drying cost.

The statement above, written in 1967 by a man of forty years' experience in grain merchandising, is quoted in detail to show that there is more to grain drying than merely pouring moist grain into the intake hopper of a dryer and removing dry grain from the discharge hopper; it is not a business for dullards. Old-time horse-trading was simple in comparison, and the complications cited in the quotation are by no means all those involved.

Others will be noted shortly, but first more details about the "invisible loss" or underestimated shrink, mentioned above, which tends to be a bit tricky. It has been and still is a matter of contention between farmers or others who are having grain dried, and the elevator men or others who are drying the grain. This is a product of the practice of buying and selling grain on a basis of wet weight, rather than dry weight, so that after the grain is dried the basis for calculating the moisture content is a lower total weight than before the grain was dried. An example will illustrate this.

Suppose one has 100 pounds of corn, with 20% moisture content, wet weight basis, and wishes to have this dried to 15.5%. At first glance it seems reasonable to suppose that 4.5 pounds of water must be removed from the 100 pounds of corn, to achieve a moisture content of 15.5%, since 20 minus 4.5 equals 15.5%.

Not so. Before the grain is dried, there are 80 pounds of dry matter and 20 pounds of water, giving a total wet weight of 100 pounds. When the grain is dried, the new basis for calculating the moisture content is not 100 pounds, but the total weight of the corn when dried to 15.5% moisture content. This total weight turns out to be 94.678 pounds, with 80 pounds of dry matter, as before, and 14.678 pounds of water; 14.678 divided by 94.678 equals 15.5%. Since the total weight now is 94.678 pounds, 5.322 pounds of water have been removed, not 4.5 pounds. Tables have been constructed to enable one to see this at a glance.

The aim in drying grain, then, is not to remove as much water as possible, but to remove as little as possible to meet a given grade, or to make the grain safe for storage for a given length of time at a given temperature, and at the same time to cause as little damage to the grain as possible and, over all, to do this at as low a cost as possible. To achieve these ideal but varied aims never is easy, and sometimes is impossible, for reasons now to be discussed.

Grain can be dried at too high a temperature, and to too low a moisture content, either of which results in loss of quality or cash, or both.

If corn is dried at temperatures above 140° F. (60° C.) the endosperm of the kernels is chemically altered. This may not reduce the value of the grain for feeding, but does reduce its value for the extraction of starch and protein, because some of the protein will end up in the starch, which reduces the grade of the starch, and at the same time results in loss of high-priced protein.

For this reason the corn-processing industries do not want grain that has been dried at too high a temperature. They or their buyers cannot, of course, determine the temperature at which the corn they purchase has been dried, but they can and do test germinability of the corn. Reduction in germinability from drying at too high a temperature occurs at about the same temperature that results in the chemical changes that make separation of starch and protein difficult. So the corn processors prefer corn of high germinability. Factors other than high temperature may result in loss of germinability, such as invasion by fungi (corn processors do not want grain heavily invaded by fungi, either) and mechanical damage. But in any case corn of high germinability is preferred to corn of low germinability, even though the processors will not pay a higher price for such corn. That is, premium quality in this case does not command a premium price, which is one of the facts of life of grain merchandising.

The value of corn for feed is not affected by temperatures up to 160° F., and so if the grain is to be used for feed it should be possible to dry it at a higher temperature than if it is to be processed for extraction of starch and protein. From the standpoint of efficiency in drying, a higher temperature is desirable, because it costs less per pound of water removed. If grain handlers could determine beforehand to what use the grain was to be put, the problem would be simplified. Theoretically this is possible, of course, but commercial grain handlers do not usually know beforehand the ultimate destination and use of the grain they handle.

Drying to too low a moisture content can also result in difficulties, financial and otherwise. As the moisture content of the grain is reduced, the removal of more water becomes more expensive. Also any reduction of moisture content below 15.5% means driving off water worth 2¢ per pound, as mentioned above. In some cases, as for long-term storage or for long transit in shipment, it may be desirable or necessary to reduce the moisture content to 13.5%. If the moisture content is reduced to below 13.0%–13.5%, stress cracks and fissures develop in the endosperm of the seeds, and such grain is highly susceptible to breakage as it comes from the dryer and in subsequent handling. Broken material is undesirable on several counts, and it constitutes a factor in grading and therefore in the price of the corn.

Some of the hazards of mixing lots of different moisture content were discussed in Chapter 5, but a further word may be in order here. If lots of 12% and 18% moisture content are mixed in equal amounts, the average moisture content of the mixed lot will be 15.0%, but the keeping qualities of at least half the mixture will be more like grain of 18.0% moisture content. The moisture meter may indicate that such grain is safe for storage, whereas it actually is grain of high storage risk. If the lot originally at 18.0% moisture content had been invaded by storage fungi before mixing, the storage risk of the mixed lot would be even higher. If an elevator receives a lot of corn of which half has been dried, and the other half not, and the average moisture content is still too high for safe storage, the operator has an impossible task. If the grain is dried again, as it almost has to be, the previously dried portion will become even drier and probably will produce in subsequent handling a great many broken kernels, resulting in decrease in grade and price; the previously moist portion may not be reduced in moisture content enough to be safe for storage, and so may be invaded by fungi and spoiled. No method is now available for the detection of the range in moisture content in such mixed lots, although work is under way to enable this to be done.

Drying of corn so as to maintain its quality, including high germinability, to avoid breakage in subsequent handling, and to achieve a low enough moisture content that will permit storage and shipment without spoilage, is becoming more and more important in national and international trading. Obviously this is not a simple procedure, and involves considerable technical knowledge, not only of drying procedures but also of economics and of the factors that affect corn quality even if not all of these enter into grade determination. Since a major benefit to be derived from the drying of grains and seeds is protection from invasion by fungi, it is desirable for those involved to have some knowledge of fungi. Some do, some do not.

In the developing countries the drying of grains and seeds so as to maintain high quality for food and feed is accompanied by so many difficulties that the problem seems almost hopeless of solution. How can a small farmer near Veracruz, Mexico, for example, dry his corn when it is raining nearly every day, and, when not raining, the weather is hot and humid, everything is being overgrown by fungi, and grain-infesting insects abound. How to help him? He may be growing some of the improved varieties of corn, but if the grain is being consumed by insects and fungi the benefits he gets from the increased production of improved varieties are minimal. The poor in many of the tropical and subtropical regions of the world subsist in part, and sometimes chiefly, on moldy grain, and some of these fungi growing in the grains and seeds produce injurious or lethal toxins. For these people, the preservation of quality in stored grains and seeds is a problem of direct and vital importance, and one which will not be solved by high-level discussions. Up to now very little research, extension effort, and education have been devoted to this aspect of storage.

AERATION

The chief functions of aeration systems for stored grains are (1) to reduce the temperature of the grain — at 40° to 50° F. (5° to 10° C.) storage fungi grow very slowly, and insects and mites are dormant; (2) to maintain a uniform temperature throughout the mass or bulk, which prevents transfer of moisture. Aeration does this much more cheaply and effectively than does transfer of grain from bin to bin, and aeration does not result in any increase of cracked and broken kernels. Unquestionably, the development of the principles and practices of aeration constitutes a major advance in the art and science of maintaining quality in stored grains.

The idea of aerating moist stored grain to help reduce or prevent spoilage is not new, but there could be no effective aeration of grains stored in bulk until electric power became common and cheap. Kelly (91), some of whose work was mentioned in Chapter 5, in the late 1930's was the first to test and compare methods of aerating grains by means of cowls turned into the wind and connected with ducts running through the grain, and by means of forced ventilation by fans driven by electric motors, and to get data on the condition of the grain before, during, and after aeration. He was interested mainly in reducing the moisture content of the grain by ventilation, and although he measured temperature of the grain during ventilation, the maintenance of low and uniform temperature throughout the grain was not his chief goal. A couple of his statements are of interest.

He commented: "The wheat in the unventilated check bin F-3 began to increase in temperature, *at a point near the center and 1 foot below the upper wheat surface* [italics ours] immediately after being placed in storage." The reason why deterioration began in that particular location will be seen from the discussion of air circulation within bulks of grain, below.

He went on: "The grain in bin F-5 (perforated walls and floor), from the same lot as that in bin F-3, also began to rise in temperature near the center of the bin soon after being placed in storage. When this temperature reached 105° F, after 26 days in storage, the wheat was moved for cooling and returned to the same bin for further observation. When moved, the wheat in about *one-fifth of the bin near the top center* [italics ours] was found to be musty and had a moisture content of 16.1 percent. However, after thoroughly mixing with the balance of the wheat, the odor of the musty portion was not strong enough to be detected." Again, spoilage occurred in the center of the grain, near the top. That the grain, after the musty and fungus-invaded portion had been mixed with the sound or less musty portion, had no detectable musty odor means that so far as inspection was concerned it would pass as sound grain. Whether it was wholesome or not is another matter. Such grain we now know would be of very high storage risk when sold to a terminal elevator or mill, and we now have methods available for detecting just such lots of high storage risk.

In the 1940's and early 1950's aeration of grain by forced movement of air through the grain mass was the subject of much investigation by agricultural engineers in experiment stations and in grain storage firms. As a result of their work rapid progress was made in understanding the

115

conditions that prevail in bins of stored grains, and the changes that go on. Before this work was done grain whose moisture content was so high as to make the maintenance of its quality in storage doubtful, and grain in which incipient or advanced spoilage was detected by a rise in temperature somewhere within the mass, would be emptied out of the bin and run into another bin. This aerated the grain after a fashion, mixed it some, and broke up hot spots. Sometimes this stopped the trouble, sometimes it just distributed the trouble through the whole mass. This transfer was expensive in itself, and it meant that a bin or bins always had to be kept empty to permit such transfer. Whether it is being used or not, storage space is costly. To see why aeration not only reduced the necessity for such turning of grain to maintain its condition, but also was much more effective than turning in maintaining good condition of the grain, it is necessary to explain something of what happens in grain stored in bulk.

Grains such as wheat, barley, rice, sorghum, and shelled corn stored in mass have a void space of 40%–45% of the volume occupied by the grain. This can be determined very easily by filling a graduate cylinder or a quart measure with grain, then pouring in some liquid, such as oil, which is not absorbed by the grain itself, and measuring the amount of liquid necessary to fill the measure up to the surface of the grain. If the grain contains many broken kernels and much fine material, or "fines," this may pack into the spaces between the kernels and reduce the air space. As mentioned in Chapter 2, when bins are filled by means of a spout that empties into one place in the bin, this fine material will accumulate in a column below the spout — referred to as the "spout line" — and so result in an almost solid column of material through which air passes with difficulty.

As described in Chapter 7, grains of moisture contents normally encountered in storage have a low specific heat and high insulating properties. This means that if there are temperature differences between different parts of the grain in a given bin when the bin is filled, these are not equalized quickly by conduction of heat (conduction is used here in its technical sense, as distinguished from convection or radiation) from one portion to another. If such temperature differences exist when the bin is filled, or if they develop later, as when the grain at the surface and next to the walls is chilled in winter, air currents are set up in the mass. Differences in moisture content of different portions of the grain also result in circulation of air.

Concerning this air circulation and resulting transfer of moisture in stored grains Johnson (89) wrote:

Shelled corn and wheat have a void space of about 40% of the total volume; the interstitial air assumes a specific weight which is dependent on temperature and humidity for a given atmospheric pressure. At normal grain-storage moisture contents [he is referring to grains with 13.5%–14.5% moisture content, wet weight basis] the air relative humidity is in the range of 65–75%, and there is very little variation in specific weight over this range. However, temperature differences of 40 to 50 F are found to result in significant differences in the specific weight of air. For example, air at 85 F and 75% relative humidity has a specific weight of 0.07 lb per cu ft at standard atmospheric pressure, while air at 45 F and 75% relative humidity weighs 0.078 lb per cu ft, or 11% more.

It is of interest to calculate the amount of moisture which could be transported in a hypothetical case, assuming a sharply defined temperature distribution and a common shape for the grain mass . . . The bin is 120 ft in diameter and 30 ft high. The difference in specific weight of the interstitial air which results from the difference in temperature would produce a pressure of 0.017 in. of water for movement of air along flow path A, a distance of 90 ft. An extrapolation of the data of Shedd for air flow in grain indicates an air flow of 0.05 cfm per sq ft through clean shelled corn.

The amount of moisture released at the cool top surface of the warm part of the grain mass may be calculated from equilibrium relative humidity data and the psychrometric chart with the assumption that the convected air attains thermal and moisture equilibrium with 14.5% moisture corn at 85 F before being cooled to 45 F. In this case 21 days would be required to move enough water to increase the moisture content of a top layer six inches deep from 14.5 to 20%. This is recognized as a hypothetical example because of the sharp delineation between the warm and cold zones and because only one flow path is considered, but it is indicative of the magnitude of moisture migration which can result from the existence of lateral temperature differences.

The potential for convection-air movement in a grain mass depends on the existence of lateral temperature differences, the magnitude of the resulting variations in specific weight of the interstitial air, and the vertical extent of the zones of cool and warm grain. The air flow which results is dependent on the ease with which air may move through the grain mass as affected by the kind of grain, amount of foreign material, degree of packing, and moisture content. The relatively low air-flow resistance of shelled corn as compared to wheat is notable in this respect.

If the air movement produces a moisture transfer, there must be a zone of lower temperature to increase relative humidity or condense water vapor by cooling. Condensate dripping from the cold roof surfaces of bins is one manifestation of moisture transfer.

Surface-layer moisture accumulation is often called "top sweating"; this type of moisture migration is the most common because of the temperature differences which usually develop in those geographical areas which have low atmospheric temperatures during the winter. However, moisture movement will be toward any part of a grain mass which has substantially lower temperature, provided the temperature distribution is favorable for initiating and sustaining convection air currents. If there is an extensive grain surface which is relatively cold as in flat storage, the air rising from warm grain will be cooled, thus increasing its relative humidity and resulting in moisture transfer from air to grain. Since this top layer may receive both heat and moisture from the convected air, conditions soon become favorable for mold growth with further heating and moisture transfer.

This was written in 1957, and was based on work with grain stored in commercial bins, when Johnson was an engineer with the Grain Research Laboratory of Cargill, Inc. It is as sound now as it was then. Basically, the facts are that in a grain mass cold air will flow downward, and warm air upward. The warm air absorbs moisture from the grain through which it flows. As this warm air enters grain of lower temperature it is cooled and its humidity rises. Most commonly the cool grain is at the surface of the bin, but when bins are loaded with grain during the fall season or later, some truckloads or carloads of grain may be much lower in temperature than others, and so unless the grain is thoroughly mixed, as by transferring it to another bin, there may be considerable differences in temperature between one portion of the grain and another. So the grain of low temperature in *any* part of the mass may slowly or rapidly increase in moisture content. It has been assumed by some that the only significant increase in moisture content takes place where condensation occurs. This is false, as mentioned but not emphasized by Johnson when he says that "the air rising from warm grain will be cooled, thus increasing its relative humidity and resulting in moisture transfer from air to grain." This transfer of moisture *without* condensation probably is much more common and important than condensation, since it results in raising the moisture content of a portion of the grain to the point where fungi can grow, and these slowly or rapidly increase the moisture content by producing metabolic water.

Johnson, by the way, was well aware of the role of fungi in grain spoilage.

It has been claimed, as late as 1967, that spoilage of grain by fungi results in drying of grain because of the heat liberated, so that a kernel of corn undergoing spoilage by a fungus functions as a small burner. The small burner in this case is producing water as well as heat. According to the formula:

$$C_6H_{12}O_6 + 6\,O_2 \rightarrow 6\,CO_2 + 6\,H_2O + 677.2 \text{ calories}$$

180 g.   108 g.

That is, the heat produced by burning 1 gram molecular weight of sugar, 180 g., results in 108 g. of water and 677 calories. This is sufficient heat to raise the temperature of the metabolic water produced by the process only about 6.5° C. The grain being decayed in this process also has some weight, and contains water which must also be heated if it is to be driven off. Grain as it heats, either in the laboratory in insulated containers, or in a bin as a result of fungus invasion, does not *decrease* in moisture content, usually, but *increases*. This is gone into in detail in Chapter 3. A hot spot in a bin of grain may "burn itself out" if left undisturbed (it may also lead to heating and spoilage of the entire contents of the bin) and the residual heat may dry out a portion of the spoiled grain, and grain adjacent to it. This has given rise to the false notion that microbiological heating generally, or usually, or inevitably results in loss of water from the heating mass. In the final stages it may do so, depending on how the water evolved in the heating process was dissipated, but by then the grain is spoiled.

The transfer of moisture resulting from temperature differences between different portions of a bulk of stored grain is not a phenomenon limited to large masses. It will happen in a two-quart jar of grain stored in the laboratory if one side of the jar is kept constantly at a temperature several degrees higher than the other. This transfer of moisture is likely to be much more rapid in grain stored at moisture contents above about 13% (in the starchy cereal seeds) than in grain stored at 11%–12% moisture, partly because in the grain at higher moisture content there is more water available for transfer and partly because of the relatively rapid increase in specific weight of the air as the relative humidity increases, as mentioned by Johnson. For the same reasons, such transfer is likely to be much more rapid at higher temperatures than at lower.

Aeration or forced ventilation goes a long way toward avoiding transfer of moisture because a uniform temperature can be maintained throughout the bulk. Also any desired temperature can be maintained throughout the grain either by aerating when the temperature of the outside air is low, or by chilling the air artificially. The airflow required to produce a uniform temperature throughout the grain is not great — from 0.025 to 0.15 cubic feet per minute per bushel of grain.

As to its utility, Johnson stated, "No deterioration of grain quality was observed in the operation of the aeration system described above and other aeration tests have resulted in full maintenance of grain quality." This is supported by Kaufmann (90) who in 1963 wrote: "Aeration's effectiveness is demonstrated by an example from Cargill experience. We have stored several million bushels of wheat and corn for lengthy periods without grade loss. Ten years ago, without aeration, the likelihood of success in such long-term storage would have been at best questionable. Grain storage has become a far more exact science in a decade, thanks to improved drying and binning, good temperature checks, carefully planned sampling, spot treatment of infestations, and aeration. And it should be noted that these improved handling procedures have increased not only the life of the grain; they have extended the elevator superintendent's life span as well."

Although the primary purposes of aeration are to lower the temperature of the grain and to maintain a uniform temperature throughout the grain mass, it usually will result in some decrease in moisture content also, sometimes as much as 1%. Moreover, this moisture, even if removed slowly, is removed from the *outside* of the kernels, so that even if the average moisture content of the individual kernels is high enough to permit invasion by fungi, such invasion may be retarded appreciably by the lower moisture content of the outer layers of the kernels, resulting from aeration. Aeration does not lower the temperature of all the grain throughout a bin gradually and uniformly. A rather sharply delimited cold front is developed where the air enters the grain, and this front moves through the grain in the direction of air flow. Behind this cold front, all the grain has the same temperature as the air moving through it; ahead of the cold front the grain remains at the temperature it had when it was put into the bin. With wheat aerated at the rate of 0.1 cubic foot of air per minute per bushel of grain in the bin, about 120 hours are required for the front to move through, or about five days of continuous aeration. The *size* of the grain mass does not matter, since the base of calculation here is cubic feet of air per bushel of grain per unit of time; a mass of 1 bushel would require 0.1 cubic foot of air per minute, and a mass of 100,000 bushels would require 10,000 cubic feet of air per minute. The last grain to cool is that at the exhaust side of the bin, and if damage is to be avoided this grain must be cooled before spoilage begins. This must be kept in mind when estimating the likely success of aeration in preventing spoilage.

Most of the material summarized and discussed up to this point in the chapter has come from grain with moisture contents no higher than about 15.5%–16.0%. It embodies the results of some rather extensive tests in the laboratory concerning the relation of moisture content, temperature, and time to the growth of storage fungi and the resulting effects on grain quality, and also carefully monitored tests in bins of various sizes and shapes, carried out over a period of years. The grains involved were chiefly those in commerce, where maintenance of quality had to be assured until the grain reached the buyer, whether nearby or thousands of miles and some months away.

REFRIGERATION

Since about 1965 an increasing amount of work has been done with refrigeration of stored grains at high moisture contents, sometimes above 20%, and at temperatures just above or even below freezing. Reports from England (92) and the United States (93) indicate that such low-temperature storage of high-moisture grains may have real possibilities, but many problems connected with it remain to be explored. The costs of operating a refrigeration system for cooling grain appear to equal or exceed the costs of drying, and the moisture content of such chilled grain has not been reduced; if the grain is to be marketed later, it must be dried. Also, refrigeration does not eliminate all risk of spoilage. To quote again from Johnson (89):

However, there are disadvantages in cooling to a low level in that a great difference between grain and air temperature will be encountered if storage is continued through the summer with the possibility of subsequent moisture movement. If cold grain is moved out of storage during the warmer months, a significant amount of moisture may be picked up because of grain temperature being below the atmospheric dew point. This moisture would not be sufficient to increase the average moisture content appreciably, but it would occur on the kernel surfaces where it could initiate activity of the everpresent molds. *Such condition problems with cold grain moving out of storage during the summer months often have been observed* [italics ours]. Although the optimum storage temperature has not been determined with regard to these factors, the range of 40 to 60 F appears desirable for aeration cooling at terminal elevators.

Qasem and Christensen (22) reported that fungus-free corn stored for two years at 18% moisture content and 15° C. (59° F.) germinated 96%, and when stored at 15.6%–15.8% moisture content and 5°, 10°, and 15° C. (41°, 50°, and 59° F.) this fungus-free corn retained a germinability of 100% for two years. However, when inoculated with spores of a mixture of *Aspergillus* species and stored at 15.9% moisture content and 15° C. (59° F.) the germination after two years was only 48%. Also when inoculated with spores of A. *flavus* and stored at 18% moisture content and 15° C. for 4.5 months, germination was reduced to 62%. Christensen (2) found that No. 2 corn stored at 12° C. (52° F.) decreased in germination from 60% at the beginning of the test to 42% in six months, to 36% in one year, and to 1% in a year and a half.

Some fungi, when sufficient moisture is present, will grow at temperatures well below freezing. If the moisture content of starchy cereal grains, including corn, is above about 22%, bacteria can grow on and within the seeds, and if the temperature is high enough to promote rapid growth they may cause incipient spoilage within 24 hours. This may not be accompanied by any obvious discoloration or detectable odor at the time, but may lead to very rapid spoilage later. Very definitely, more work is needed on the microbiology of high-moisture-content grains stored at low temperatures. To properly evaluate the use of refrigerated storage of high-moisture-content grains, we must learn more about the conditions under which such storage is *not* effective as well as the conditions under which it *is* effective.

9 *INSECTS, MITES, AND RODENTS*

INSECTS

Nature and extent of losses. Insects take a terrific toll of stored grains and seeds; according to an FAO report of 1948 (99), in 1947 the losses of breadgrains and rice throughout the world totaled about 33 million tons, enough to keep 150 million people alive for a year. It was estimated that at least 50% of this loss was due to insects. Such losses are not new, by any means, and their magnitude probably is much less now, as a percentage of the harvested crop, than was the case before control measures became an integral part of grain handling and storage in the more developed countries. According to a brief account in the *Transactions of the British Entomological Society of London* in 1870 (106), "Mr. J. Jenner Weir . . . exhibited specimens of grain attacked by weevils: from 74 tons of Spanish wheat 100 cwt. of weevils had been screened, and these must have consumed several times their own weight of grain before arriving at maturity: in August, 1868, some American maize was stored, weighing 145 tons; in August, 1869, this was found to be infested with weevils, and 6 cwt. of the beetles were screened out; in December 29 cwt. more were screened out, making a ton and three-quarters in all." This was calculated to amount to 4,056,729,600 adult weevils. The grain by that time probably contained at least an equal or greater weight of larvae and pupae within the kernels, since the insect was identified as the rice weevil, *Calandra oryzae* (now *Sitophilus oryzae*), which spends its larval and pupal stages within the kernels. Excreta of the insects probably amounted to at least several times that amount. Since the rice weevil is a relatively strong flier, this infestation probably spread to many nearby stocks of grain. How the remains of this heavily infested grain were disposed of was not mentioned, and no special concern was expressed about the heavy infestation. It seems probable that moderate to heavy loss from insect infestation was accepted as one of the risks of storage and handling of grain at that time. In looking through the volumes of this journal from 1834 to 1900, we could find only the quotation above dealing with grain-infesting insects, from which we may assume that infestations of lesser degree were so common as not to merit much notice.

Cotton (95) in 1950 stated that insects in stored grains and milled products caused a loss of at least $300,000,000 annually in the United States. This included cost of control, which of course adds to the cost of the final product.

In many of the subtropical and tropical countries, where some of the grain-infesting insects occasionally or commonly invade grains in the fields before harvest, and where the temperature is favorable for rapid multiplication of the insects throughout the year, and the relative humidity is high enough through much of the year to keep grains and seeds moist enough for invasion, losses may be much higher than in the more temperate or colder countries. We have seen corn stored in bags in a warehouse in Veracruz, Mexico, where within a few months after harvest the bags, and the walls and floors of the warehouse, were covered with crawling weevils.

The major economic loss caused by grain-infesting insects is not always the actual material they consume; to this must be added the amount contaminated by them and their excreta and made unfit for human consumption. Individuals and societies differ in their opinions of fitness or unfitness for consumption of various ingredients in the foods they consume, but certainly most people do not willingly accept or relish weevils or moths, whole or ground up, or the excreta of these insects, in their food. Partly this may be for esthetic reasons, to some extent an educated and artificial revulsion toward "filth"; but recently it has been found that some grain-infesting insects harbor in their gut potentially harmful bacteria, such as pathogenic *Salmonella* (a common cause of food poisoning, found in feces), hemolytic *Streptococcus,* and *Escherichia coli* (also from feces), and they may well harbor also viruses capable of infecting man or his domestic animals. That is, insects in grains and grain products are undesirable not only from an esthetic standpoint, at least to most of us, but from a public health standpoint as well. The development by workers in the Food and Drug Administration (FDA) of techniques for the detection of insects, insect fragments, and insect excreta in grains and grain products, and the establishment, in the 1950's, of strict standards concerning such contamination, have resulted in a great reduction in infestation of grain stocks in the United States. There has also been a reduction of insect infestation of products imported from foreign countries, since these must meet the standards established by the FDA, and in this way the actions of the United States have had a beneficial effect far beyond its borders.

We have not seen any recent estimates of the amounts of stored grains and grain products damaged or destroyed by insects in the United States. However, in the many thousands of samples of wheat, corn, and other grains from commercial elevators and warehouses throughout the country that we have examined over the past fifteen years, evidence of insect damage has been uncommon, and a living insect an extreme rarity. For example, in about 500 samples of hard winter wheat from storehouses of several grain firms examined over a period of five years, not a single live insect was found. In about 400 samples of corn of grade No. 2 from grain being loaded into ships from New Orleans to Baltimore, for export, all of which were examined thoroughly, the only insect of any kind found was one living adult sawtoothed grain beetle. The freedom of commercially stored grains from insects and insect damage in the United States is the best possible proof of the effectiveness of the program of insect control.

As research turns up more information concerning the microflora associated with grain-infesting insects, and as means become available for precise determination of the number and kinds of microflora in the products of grains and seeds, it is likely that the standards of purity will be raised even higher, and not only for food products, but for feeds as well. In some of the economically underdeveloped countries, the establishment of high standards of purity for foods and the reduction of contamination by insects and rodents will constitute a very difficult problem for many years to come. This is especially true where few or no facilities are available for insect-free storage of grains, even those grains that are imported free of insects, and where there are few facilities for fumigation or otherwise ridding the grains of insects while they are awaiting processing for consumption. In many lands, for the small farmer who depends for food on the small stock of grain of his own harvest, stored in a pile in the corner of a room or in bags exposed to all sorts of infestation, the problem of pure food is strictly academic, but it need not necessarily remain so. Education, research, and extension can help in this as in other aspects of the economy.

Some of the common grain-infesting insects, especially the weevils, the larval stage of which develops within the kernel, regularly carry with them, outside and in, a rather heavy load of storage fungi. As these insects develop in stored grain, they increase the moisture content of the grain, through water given off in their respiration; in one test (14) they increased the moisture content of the infested portion of grain over 5% in two months. So if these insects are able to establish themselves in grain whose moisture content is just below the level where storage fungi can begin to grow, within a short time the conditions may become favorable for rapid development of fungi, with resulting development of musty and decayed grain. The insects also raise the temperature of grain they infest, which means more rapid development of the fungi. The insects themselves cannot raise the temperature above a maximum of about 40° C. (104° F.), and many of them cannot endure a temperature as high as that. The fungi, however, with optimum growing conditions, can raise the temperature up to 55° C. (131° F.), by which time the grain is totally spoiled. Heating may sometimes proceed to spontaneous combustion, with total loss of the grain and sometimes of the structure in which it was stored.

Major kinds of grain-infesting insects, and their habits. More than fifty species of insects infest grains and grain products, but of these only about a dozen are major pests of stored grains and seeds. Some of these are much more prevalent in some countries than others; they are especially common in the tropics, their native home. Most are now worldwide in distribution, having been carried from region to region and country to country by commerce. Some of them are rather strictly adapted to a diet of seeds, and even to certain kinds of seeds, while others are scavengers that can live on a great variety of animal and vegetable products.

The most widespread and destructive of these pests are the beetles and moths. Among the beetles are weevils (characterized by long snouts), principally the granary weevil, *Sitophilus granarius*, and the rice weevil, *S. oryzae*; dermestid beetles, of which the khapra beetle, described in some detail below, is an example; various grain borers, including the lesser grain borer, *Rhyzopertha dominica*, and the saw-toothed grain beetle, *Oryzaephilus surinamensis*. The moths include the

Angoumois grain moth, *Sitotroga cerealella* ("Angoumois" is from the province in France where severe damage by this insect was first observed in the early 1700's), and the Indian meal moth, *Plodia interpunctella.*

Adult females of the granary and rice weevils hollow out a minute cavity in a sound seed, usually near the embryo, deposit an egg in this, and close the hole with a plug of gelatinous material. Under favorable circumstances the adults live for several months and each female may deposit several hundred eggs. The egg hatches into a larva, the larva consumes the interior of the seed, and pupates within the hollow it has made. Under favorable circumstances the life cycle is completed in 25–30 days. The granary weevil cannot fly, is distributed chiefly by shipments of infested grain or other materials, or by means of trucks, railway cars, ships, planes, or other conveyances that previously carried infested materials, and is pretty strictly a pest of stored grains and seeds. In spite of its inability to fly, once it has reached a given grain elevator or storage site it seems to be able to travel fairly fast throughout the premises. The rice weevil is a comparatively strong flier, and so can travel from granaries to fields of ripening grain and lay its eggs in the seeds, and such grain may have a hidden infestation when it is brought to the warehouse. In the southern part of the United States, in Mexico, and in other warm regions of the world where grains are grown, infestations of corn and sometimes of other grains in the fields by this beetle are not at all uncommon.

The grain borers, in general, lay eggs on the grain or in the fine material that accompanies it, and the larvae feed on the grain from without or within. Some of these consume the kernels almost totally. A good example is the lesser grain borer (named "lesser" because of its small size, not because of its appetite, which is enormous), one of the most damaging insect pests of stored grains in the United States. It is capable of relatively long flight, and under favorable circumstances it can spread rapidly over considerable distances; it probably also can be carried by heavy winds. In traps consisting of a net at the end of a revolving boom, operated at Hutchinson, Kansas, in the summer of 1942 (97), 256 lesser grain borers were caught in 346 hours of operating time. Each trap sampled only a very small portion of the air (the size of the net was not given) but the evidence indicated that adults of this insect must have been relatively common in the air at that time and place — near a large grain storage site. In the same time, only 9 rice weevils were caught, although they also are capable of rather sustained flight.

The dermestid beetles are scavengers, and consume a large assortment of plant and animal products, including stored grains. The most widely distributed dermestid in the United States is the black carpet beetle, *Attagenus piceus*, but it is not a serious pest. Several species of *Trogoderma*, also dermestid beetles, attack stored grains and grain products; the most destructive of these is *T. granarium*, the khapra beetle, which deserves a fuller account.

The khapra beetle is a native of India, Ceylon, and Malaya, and got its name in India from its habit of congregating in the cracks and crevices of buildings made of bricks known as Khapra. It is considered to be the world's worst pest of stored grains. It can consume and live on many kinds of grains and seeds, grain products such as flour and spaghetti, tapioca, nut meats, dried milk, animal products, including hides, blood meal, bone meal, bones, animal glue, fishmeal, dried fish, and almost anything of organic nature.

The newly hatched larva is about 1/25 inch long, the mature larva ¼ inch long. Young larvae are unable to attack sound, unbroken kernels of grain, but thrive on fragments until they have reached the 4th instar, when they are able to chew into whole kernels. When so feeding, the mature larvae leave many fragments on which the young larvae can feed. The adults are able to feed on whole kernels. Between larvae and adults, this beetle can cause total destruction of the stored grain and other materials that it attacks.

This destructive insect pest of stored grains has been transported to and become established in Africa, Asia, Australia, and Europe. The danger of its introduction into the United States was recognized by officials in the Plant Pest Control Division of the USDA many years ago, and the insect was intercepted many times at ports of entry, and excluded.

It was imported into California during or shortly after World War II, presumably in burlap, fishmeal, or pistachio nuts from India (107). When first encountered in rather large numbers in a California warehouse it was mistaken for the domestic carpet beetle, to which it is closely related and from which it can be distinguished only by an expert, and the measures taken to control it were ineffective. By 1949, the 300 tons of grain in the warehouse where it had been discovered were a total loss.

In 1954 a survey was begun of storage places of all types as well as of mills and feed-mixing plants, to determine the extent of the infestation. Some feed-mixing or feed-manufacturing plants use many ingredients, from different countries, and some of these ingredients come in used bags or sacks which may harbor the beetle. One dealer whose warehouses contained khapra beetles imported materials from firms in thirty states of the United States and from three states in Mexico, and the feeds and feed grains from this plant were in turn shipped to many places, primarily in western United States.

The number of insects that developed in some bins of infested grains was astounding. At one warehouse, larvae dropped over the walls in such numbers that when vacuumed up at the end of a week they filled a fifty-gallon drum. In another warehouse enough larvae were collected to fill a half-ton pickup truck. The larvae also are amazingly tough; kept at 90° F. they have lived for more than five months without food, and at lower temperatures they have been reported to survive for three years without eating. Fortunately, neither larvae nor adults can travel far under their own power — the insect is distributed by man in the shipment of goods.

In 1954, the khapra beetle was found in 25 warehouses, in California, Arizona, and New Mexico; and in 1955 225 infested premises were found, 158 in California, 63 in Arizona, and 4 in New Mexico. Even to men in the Plant Pest Control Division, the task of eliminating this beetle from the country, with so many infestations, must have seemed formidable, if not almost hopeless.

Infested premises were quarantined and fumigated. For fumigation it was necessary to cover entire buildings with gas-tight plastic sheeting (because of the habit of the larvae of crawling into cracks and crevices of walls, etc.), which required the development of new techniques. According to the situation report issued in 1955 (107), "One warehouse yet to be fumigated has a capacity of 2.5 million cubic feet, and has elevators rising to 125 feet. 'Tarping' such a building at the rate of about $8 per thousand cubic feet and inserting more than 6 tons of methyl bromide is not an inexpensive process." The fumigation was accompanied by an intensive cleanup program — probably the first in the operating lifetime of some of the warehousemen — to rid the premises of all materials and sites that might harbor the insect. The sanitation program was extended to uninfested warehouses and mills too, and operators who previously had viewed cleanliness in their warehouses and mills as an unnecessary and expensive refinement for the first time began to realize its value, with beneficial results in the control of other pests as well. At that time (1955) those in charge of the elimination of this pest stated that ten years should suffice to judge the success of the program. This turned out to be a remarkably accurate prediction. According to the report of April 23, 1965 (108), during 1964 only eight infestations of the khapra beetle were found in the United States, all in Arizona. "These included such diversified operations as a hog farm, a chicken ranch, a State experiment farm and a feed mill. Before that, the last infestations were found in Arizona in July, 1962, in California in January, 1962, in New Mexico in May, 1959, and in Texas in April, 1960." In 1965, 13,903 properties in 775 counties in 28 states were inspected, and no khapra beetle infestations were found (110). According to the USDA *Yearbook of Agriculture, 1966* (94), "By 1965 there were no known infestations [of the khapra beetle] on the North American Continent, despite a continuing record number of interceptions at many ports of entry in the United States, Canada, and Mexico." It has been found a few times since and probably will be again, but it seems to be under control.

The record of interceptions at ports of entry makes the exclusion of this beetle seem all the more remarkable. In 1964, khapra beetles were intercepted 368 times at ports of entry (109), including 2 interceptions at Dulles International Airport. Slabodnik (104) mentions that opening the St. Lawrence Seaway led to the possibility of introducing the khapra beetle directly into the midwestern grain storage and milling establishments. This was not just theory; in ships arriving at Seaway ports, cargoes contaminated by the beetle included sacks and bags, baled rubber, steel wire, cotton waste, automobiles, and soiled linen, and larvae of the beetle were found feeding on glue between the layers of the walls of laminated cardboard cartons. Costs of fumigating the holds of a single ship amounted to $7000–$10,000. The Plant Pest Control Division of the USDA deserves tremendous credit for having rid the country of this serious pest once it was established, and for having kept it from becoming established since, in the face of seemingly insurmountable difficulties.

Several genera and numerous species of moths are occasional and sometimes serious pests of stored grains, although usually their damage is restricted to exposed surfaces, such as the top of grain stored in bulk and the outer portion of grain stored in sacks. The adults do not feed. The females lay their eggs on the grain or on the outside of bags, the eggs hatch into very minute larvae capable of working their way through the weaving of bags to feed on the grain. The larvae feed on the germs or consume the entire kernels, and in the process produce large amounts of excreta. The larvae of some species secrete a fine, sticky webbing that fastens together the material in which they are feeding. Normally, moths are not difficult to control, although if not controlled they are capable of causing considerable damage.

Conditions that favor development of insects in stored grains. Grain-infesting insects, like grain-invading fungi, require food, oxygen, a favorable temperature, and a suitable moisture content. Some, as already indicated, do much better if the grain contains an appreciable amount of fine material, and some cannot become established unless such material is present. Since most of them are of tropical or subtropical origin, they are adapted to a fairly high temperature, 30°–32° C. (85°–90° F.). Their growth and reproduction are greatly reduced at temperatures below about 20° C. (70° F.) and many of them cease to develop at temperatures below about 10° C. (50° F.). If the grain is maintained at temperatures below this, they die. That is, maintenance of a low temperature in grain stocks, now possible with aeration, as described in Chapter 8, will greatly reduce or eliminate most insect problems. Moisture content, temperature, and time operate together in the development of grain-infesting insects as they do with fungi; with a higher moisture content a lower temperature is required to halt all insect activity, or a longer time at a given temperature. Once an infestation has begun to develop, even a small one localized in a bulk of grain, the insects increase the moisture content and temperature through their own activities, and so provide and maintain an environment favorable for their continued development.

Control. From what has been said concerning the habits and requirements of grain-infesting insects, effective control measures depend primarily on making the environment unfavorable for their development. Fumigation with poisonous chemicals is by no means the only answer, or even the best one in most cases. Control begins with sanitation, which includes the construction of insect-proof storage places — tight bins or warehouses to which insects do not have ready access at all times. Naturally, warehouses and bins must be open to receive and discharge grain, and insects can and commonly do enter with the grain; but storage places should not have holes or cracks or open windows where insects can enter at any time. The bins and warehouses should be kept clean — thoroughly and scrupulously clean. When bins are emptied they should be cleaned out — not just shoveled out, but cleaned out, with all debris, caked grain, and rubbish removed. If the grain loaded out of a bin has been infested with insects, the walls and floors of the bins should be sprayed with a lasting insecticide. Once a bin is filled with new grain, if it is to be held for any time, it should be fumigated within a few weeks, to eliminate any infestation that may be present. To wait until an infestation has developed, before fumigating, means that some grain already has been damaged and destroyed, and even if the insects are killed their activities may have provided conditions favorable for growth of storage fungi; the fungi are not likely to be killed by fumigation and may continue to develop. If the grain is to be kept for any length of time it should be at a safe moisture content when stored. Aeration, to reduce the temperature throughout the grain and to keep it uniform and low, will reduce the possibility of invasion by insects and fungi, and will go far toward retarding their development if they are present. The areas around and near the storage site should also be kept clean and free of spilled grain or debris where insects might breed or rodents find harborage. Good housekeeping, in other words, includes the outside as well as the inside wherever grains and seeds are stored and in whatever quantity — on farms, in homes, or in small or large elevators or terminals. Grains and seeds are perishable products, and must be well cared for if their quality is to be maintained.

Cotton *et al.* (97) stated, "The proper utilization of nature supplemented by artificial methods can be depended upon to conserve grain in storage for an indefinite period without material loss from insect attacks." And "In North Dakota wheat stored in steel bins and given only an initial fumigation remained free of insect damage for five years. The temperature of the grain never reached a high enough level to support insect development. In portions of the North Central States with similar climates, insect-free, low-moisture grain placed in clean bins should keep in excellent condition without further attention. It is essential to use bins that are insect-free and weatherproof. Wheat stored in south-central Kansas if untreated will gradually be destroyed as a result of insect attack."

In summary, they said:

For the entire region it is recommended that, for the conservation of wheat and other small grains, producers:

1. Store only grain that is dry, preferably with not more than 12 percent of moisture.

2. Store in weather-tight, rodent-proof bins, preferably of steel.

3. Clean out all bins before loading with grain; spray walls and floors of wooden bins and around door frames of metal bins with a residual spray.

4. Clean up and dispose of litter, waste grain, and feed that has accumulated in and around farm buildings.

5. Fumigate in two to four weeks after placing grain in storage.

6. Inspect frequently and refumigate if an infestation is discovered.

N. H. Williams, director of sanitation, Grain Division, Cargill, Inc., in 1967 stated (111): "To me plant sanitation is composed of a series of rather simple and uncomplicated items that can be readily grasped and understood. Before we attempt to master all the complicated and learned information available on the subject, we need to learn and pursue the simple and basic items that build a sound foundation."

He then listed the following ingredients of a sound sanitation program:

1. Good inbound inspections to guard against receiving contaminated, infested or mercury treated grain. (mold checks, black light.)

2. Screen and tighten structures to deny rodents and birds entry to the plant. Equip screen doors with self-closers.

3. Clean up exterior to do away with rodent harborage. Elevate all necessary material, such as grain doors and trimming lumber 18 inches off the ground on racks. Do away with scrap, trash and rubbish. Kill all grass and weeds around plant.

4. Engage a reliable exterminator to bait and trap both interior and exterior of plant. Have him gas active rodent burrows around plant exterior. *Know what he is using!*

5. Supplement exterminator's work on a plant interior with "wind up" mouse traps. Mice present a more acute problem to me than rats.

6. Develop and maintain a good sweeping and cleaning program within the plant. Give special attention to rolls, concentrators, belt supports and draw-offs. Keep overhead clean.

7. Fog and/or spray plant interior on a regularly scheduled basis to keep plant insects at a minimum and to guard against reinfestations.

8. Be sure all plant superintendents and managers are familiar with

the required standards of plant sanitation. Hold Sanitation Workshops for plant personnel.

9. Conduct regular plant inspections by a staff member or sanitarian schooled in sanitation standards and plant operation.

10. When you build a program, stick to it! Follow up on a daily, weekly, monthly basis; and keep a record of your costs, your accomplishments and your problems.

Remember, it is man himself who holds the key to the solution, and not the method or as George Wagner [director of sanitation at Pillsbury, Inc.] said last year, "The fault in most control failures lies with the person employing the method or ingredient."

An effective sanitation program, in other words, is basically good housekeeping. This is true regardless of the kind of grain or seed stored, or where it is stored. Control of insects and associated pests and vermin may be more difficult in New Delhi, India, and Tampico or Veracruz, Mexico, than it is in Larimore, North Dakota, or Winnipeg, Canada, but the basic approaches are the same: store the grain clean and dry, in clean bins that are weather tight and rodent proof; fumigate to eliminate any living insects brought in with the grain; aerate to maintain a low and uniform temperature; take samples and inspect them so that the condition of the grain in all parts of the warehouse is known at all times. If airtight storage is available, purging the grain or flushing out the air with CO_2 and maintaining anaerobic conditions will prevent increase of insects, but the moisture content of the grain must be low enough so that anaerobic bacteria and fungi (yeasts) do not grow.

Cotton and Gray (96) emphasized that "Good housekeeping is the simplest and best preventive measure" and that grain that is cool, dry, and free from dust or broken kernels is unfavorable for the development of insects. Henderson and Christensen (100) stated: "Insect control in warehouses may vary somewhat, depending on whether the seed is stored in bulk or in bags and according to the type of storage structure. The basic principles are the same in all cases. Best results are obtained by emphasizing preventive measures." They recommended thorough cleaning of the warehouse, plus a residual spray, plus fumigation during or immediately after seed is put in storage. They summarized the essential steps as follows:

Preventing damage to seed by insects and fungi after harvest can be accomplished by the following procedures:

Harvest promptly.

134

Clean and dry seed before storage.

Clean up the storage structure and apply residual spray before bringing seed in.

Apply a protectant to seed as it is placed in storage.

Keep the seed and the warehouse cool and dry.

Apply residual sprays on a periodic preventive maintenance schedule.

Fumigate if and when necessary.

The application of these control measures requires an understanding of the problems involved and some knowledge of warehouse construction, fumigants, aeration, grain characteristics, and so on, but basically it is just common sense and cleanliness. It also is costly, but not nearly so costly as the losses likely to be incurred if preventive measures are not taken. Even in Mexico, where according to Genel (5) heavy losses have been incurred in stored grains from destructive attacks by insects and fungi, there have been essentially no losses in the relatively large stocks of wheat stored for planting — simply because the stocks intended for seed are regarded as a product of high value and are cared for accordingly. Similarly in the United States one seldom hears of severe damage from insects or fungi to malting barley in storage, even though its price per bushel is about the same as that of corn, in which heavy losses from fungus damage are common — simply because the malting barley is cared for, from harvest on, as if it were a precious and perishable product, as of course all grains and seeds are. Also, well over 10,000,000 bushels of seed corn, for planting, are stored in the United States every year (occasionally some of it is stored for several years), and one never hears of any appreciable amount of deterioration in this corn; it simply is well cared for.

MITES

Mites that infest stored grains and grain products never have been given the attention that has been devoted to insects, in part perhaps because of their small size and the lesser amount of damage they cause, compared to that of insects. But at times they can cause serious damage. Sinha (103), who has studied in detail some of the mites and mite infestations in Canada, primarily in wheat stored in the Prairie Provinces, stated: "Mites are important pests of stored grain in many parts of the world, especially where the climate is temperate and large quantities of cereal grain are grown and stored. Serious outbreaks of storage mites in countries with these characteristics are not uncommon, e.g. during

1934–40 in USSR and during 1939–42 in Canada. Although over 80 species of mites occur in stored grain only a few cause serious loss." In a heavily infested sample of grain he said that there might be 250,000 mites per 100 cc of grain, about 3 ounces; this would mean a population of about 83,000 per ounce, or 1,328,000 per pound.

According to Hughes (101), mites are about as closely or as distantly related to insects as birds are to mammals. As a group they occupy a great array of environments, and a relatively large number of species live on stored products, including stored grains and seeds, burrowing into and consuming the germ or embryo. Some of them are adapted to a relatively low temperature, and can remain active and continue to feed at a temperature of 5° C. (40° F.), well below the temperature that permits any grain-infesting insects to feed and reproduce. They thrive in wheat with a moisture content of 13.5%–15.0%, the same moisture content that permits fungi in the *Aspergillus restrictus* and *A. glaucus* groups to grow, so that the presence of mites in a sample of grain is proof that the moisture content is high enough to permit invasion by fungi also. Like the weevils, they are likely to carry a fairly heavy load of inoculum of fungus spores on the outside of and within their bodies, and to provide conditions in the grain they infest for the fungi to grow. According to Griffiths *et al.* (16):

Grain-infesting mites *Acarus siro* and *Tyrophagus castellanii* were found in some abundance in samples of commercially stored wheat, the moisture content of which ranged from 13.5 to 15.0%, which is in the range of moisture content at which fungi in the *Aspergillus glaucus* group are likely to predominate. These mites, when developing in moldy grain, picked up spores of the storage fungi and carried these spores on the outside of their bodies, in their digestive tract, and in their feces. As they entered clean grain they inoculated it heavily with spores of these fungi, and later they fed to a considerable extent upon the fungi that developed. It is likely that whenever heavy infestations of these mites are found in grain, there will be damage not only from mites feeding on the embryos of the kernels, but also from the storage fungi that accompany them.

Mites, like insects, increase the moisture content of the grain in which they are active, making conditions even more favorable for the development of fungi. Even light infestations are likely to be accompanied by some damage from fungi, and heavy infestations of mites may lead to heating and consequent severe spoilage.

As an indication of our state of knowledge, or ignorance, concerning the practical aspects of some of these grain-infesting mites, in 1966 a sample of ground pig-feed was submitted to us for the isolation of possibly toxic fungi. The feed was so heavily infested with mites that, viewed under the microscope, the particles appeared to be tumbling over one another. The mite was identified for us as *Dermatophagoides farinae*, some species of which, according to Hughes (101) are free living, and others are found on the surface of bird and mammal skins. This was the first report of this mite in the United States (it was first described by Hughes in England in 1961) and the only information about its way of life and biological significance is the following, quoted from Hughes: "Large numbers of these mites were found in poultry and pig-rearing meal near Bristol by Miss G. C. Williams. It is possible that the meal is attractive to these mites on account of its high protein content." There has been almost no work with grain-infesting mites in the United States; in England, where the climate evidently is favorable to the development of many grain- and food-infesting mites, the major portion of the work on them has been devoted to descriptive taxonomy and classification. Since grains and feeds infested with mites have a moisture content too high for safe storage, the most obvious control measure is to store grains at a moisture content low enough to prevent their development.

RODENTS

Rodents, principally several species of rats, including *Rattus norvegicus* and *R. rattus*, but at times also mice, cause heavy losses of stored grains. In 1948 it was estimated (105) that in the 1940's rats were destroying in the United States 200,000,000 bushels of grain per year, worth $300,000,000–$400,000,000 at 1947 prices, and that each rat feeding on farm-stored grain cost the farmer $4.00 per year. Rats also contribute to spoilage by gnawing through walls of bins and containers of grain, allowing the grain to spill out and the rain and insects to enter. In addition, rats disseminate a number of serious diseases of man, and these diseases are present in endemic form in all countries of the world.

In the tropics rats at times cause severe losses of grains in the field before harvest. In the Philippines, for example, in the province of Cotabato, where rats have been a perennial problem, they increased greatly in 1952 and 1953 (98). In large areas of the province corn and rice crops were almost totally destroyed. Many families, threatened with hunger, abandoned their farms and moved to other provinces (in some of which there also were heavy populations of field rats). There were 100 to 1000 rats per acre, and rice was under continuous attack from planting until harvest. Emergency measures were taken, including passage of a "rat law" that required all able-bodied citizens from sixteen to sixty years of age in all rat-infested municipalities to work two days a week in exterminating rats. Various methods of control were tried, including a bounty system (which proved ineffective, as most bounty systems to control some undesirable and many desirable kinds of animals almost always have been), driving rats into grass piles where they were killed with hydrogen cyanide and machetes (up to 17,000 per day were so killed), and poisoning, especially with poison distributed by planes. In March, April, and May of 1954 it was estimated that 25,-343,546 rats were killed in the province, mostly by poison distributed by planes. Harvest losses from rats in the province of Cotabato in 1953 had been up to 90%, whereas in 1954 they were reduced to 15% of the crop.

This may be an extreme case, but it does point up the seriousness of the rat problem in some tropical areas. In places in India it is necessary to surround research fields, where crop varieties are being grown for yield tests, with rat-proof fencing set into the ground deep enough so that the rats cannot burrow under it.

In some areas of India the problem of control of rats, in the fields or in storage bins, is complicated by religious beliefs. According to a 1967 account by Malhotra (102):

More than nine-tenths of the population in rural Rajasthan comprises of Hindus and almost nine-tenths of them are vegetarians. The *Bishnois* (a Hindu caste in the region) religiously prohibit any killing specially the deer and the *Chinkaras* would not hesitate in killing the person who does so. Rural people celebrate annual feasts for dogs. They feed the pigeons daily on a raised dais known as *"Kubutran ka Chabutra"* to be found in each village. Feeding ants with sugar and offering coconut milk to the snakes and the snake god shrines is also quite prevalent. The bulls and cows are worshipped and every household contributes towards feeding the old, crippled and diseased cattle. The worship of rats is also found in Rajasthan and there is a beautiful temple of rats at Deshnouk (near Bikaner) where at least 80 kg. of grains are fed to them daily. . . .

It has been concluded that even though outright killing and immediate control of some of the harmful pests and uneconomic animals may appear fully scientific, it may be essential to find out a "balanced approach" based on gradual process in view of the deeply ingrained beliefs and religious sentiments of the people towards protection of all life.

In the more economically developed countries and in the temperate regions rodent populations cannot be eliminated, but they can be reduced and kept at a minimum, especially in and near grain warehouses. Various sorts of rat-proof constructions have been developed, and even old buildings usually can be rat-proofed at relatively low cost. In control of rodents, as of insects, cleanliness and good housekeeping are the essence; infestations on and near the premises must be kept to a minimum by constant trapping and poisoning. Where heavy populations of rats abound throughout a community or district, this control, to be effective, may have to be community-wide.

References

REFERENCES

Chapter 1. The Problem of Losses in Stored Grains

1. Bailey, J. E. 1967. Drying and cooling grain. *Cargill Crop Bulletin* 42(6):7–9.
2. Black, O. F., and C. L. Alsberg. 1910. *The Determination of the Deterioration of Maize, with Incidental Reference to Pellagra*. USDA Bureau of Plant Industry Bulletin 199.
3. Boerner, E. G. 1919. *Factors Influencing the Carrying Qualities of American Export Corn*. USDA Bulletin 764.
4. Genel, M. R. 1966. *Almacenamiento y Conservación de Granos y Semillas*. Compania Editorial Continental, S.A., Calzada de Tlalpan Num. 4620, Mexico 22, D.F. Mexico.
5. Shanahan, J. D., C. E. Leighly, and E. G. Boerner. 1910. *American Export Corn (Maize) in Europe*. USDA Bureau of Plant Industry Circular 55.
6. USDA. 1967. *Agricultural Statistics*. U.S. Government Printing Office, Washington, D.C. 20402.
7. USDA Agricultural Marketing Service, Grain Division. 1964. *Official Grain Standards of the United States*. Revised. U.S. Government Printing Office, Washington, D.C. 20402.

Chapter 2. Characteristics of Field and Storage Fungi

8. Agrawal, N. S., C. M. Christensen, and A. C. Hodson. 1957. Grain storage fungi associated with the granary weevil. *Journal of Economic Entomology* 50:659–663.
9. Amos, A. J. 1948. Moisture content of mite-infested foodstuffs. *Analyst* 73:678.
10. Christensen, C. M. 1951. *The Molds and Man*. University of Minnesota Press, 2037 University Avenue S.E., Minneapolis, Minn. 55455. 3rd ed., 1965, pp. 131–138, has a brief discussion of fungi and stored grains.
11. Christensen, C. M. 1951. Fungi on and in wheat seed. *Cereal Chemistry* 28:408–415.
12. Christensen, C. M. 1955. Grain Storage Studies 18: Mold invasion of wheat stored for sixteen months at moisture contents below 15 percent. *Cereal Chemistry* 32:107–116.
13. Christensen, C. M., and A. C. Hodson. 1960. Development of granary weevils and storage fungi in columns of wheat, II. *Journal of Economic Entomology* 53:375–380.
14. Christensen, J. J. 1963. Longevity of fungi in barley kernels. *Plant Disease Reporter* 47:639–642.

15. Gray, W. D. 1959. *The Relation of Fungi to Human Affairs.* Henry Holt and Co., 383 Madison Avenue, New York 10017.
16. Griffiths, D. A., A. C. Hodson, and C. M. Christensen. 1959. Grain storage fungi associated with mites. *Journal of Economic Entomology* 52:514–518.
17. Kaufmann, H. H. 1959. Fungus infection of grain upon arrival at terminal elevators. *Cereal Science Today* 4:13–15.
18. Lutey, R. W., and C. M. Christensen. 1963. Influence of moisture content, temperature, and length of storage period upon survival of fungi in barley kernels. *Phytopathology* 53:713–717.
19. Misra, C. P., C. M. Christensen, and A. C. Hodson. 1961. The angoumois grain moth, *Sitotroga cerealella,* and storage fungi. *Journal of Economic Entomology* 54:1032–1033.
20. Papavizas, G. C., and C. M. Christensen. 1958. Grain Storage Studies 26: Fungus invasion and deterioration of wheats stored at low temperatures and moisture contents of 15 to 18%. *Cereal Chemistry* 35:27–34.
21. Qasem, S. A., and C. M. Christensen. 1958. Influence of moisture content, temperature, and time on the deterioration of stored corn by fungi. *Phytopathology* 48:544–549.
22. Qasem, S. A., and C. M. Christensen. 1960. Influence of various factors on the deterioration of stored corn by fungi. *Phytopathology* 50:703–709.
23. Sinha, R. N. 1961. Insects and mites associated with hot spots in farm stored grain. *Canadian Entomologist* 93:609–621.
24. Tuite, J. F. 1959. Low incidence of storage molds in freshly harvested seed of soft red winter wheat. *Plant Disease Reporter* 43:470.
25. Tuite, J. F. 1961. Fungi isolated from unstored corn seed in Indiana in 1956–1958. *Plant Disease Reporter* 45:212–215.
26. Tuite, J. F., and C. M. Christensen. 1955. Grain Storage Studies 16: Influence of storage conditions upon the fungus flora of barley seed. *Cereal Chemistry* 32:1–11.
27. Tuite, J. F., and C. M. Christensen. 1957. Grain Storage Studies 23: Time of invasion of wheat seed by various species of *Aspergillus* responsible for deterioration of stored grain, and source of inoculum of these fungi. *Phytopathology* 47:265–268.
28. Tuite, J. F., and C. M. Christensen. 1957. Grain Storage Studies 24: Moisture content of wheat seed in relation to invasion of the seed by species of the *Aspergillus glaucus* group, and effect of invasion upon germination of the seed. *Phytopathology* 47:323–327.

Chapter 3. Measurement of Moisture Content

See also reference 13
29. American Association of Cereal Chemists. 1954. *Storage of Cereal Grains and Their Products.* Published by the Association, 1955 University Avenue, St. Paul, Minn. 55104.
30. Christensen, C. M., and R. F. Drescher. 1954. Grain Storage Studies 14: Changes in moisture content, germination percentage, and moldiness of wheat samples stored in different portions of bulk wheat in commercial bins. *Cereal Chemistry* 31:206–216.
31. Fairbrother, T. H. 1929. The influence of environment on the moisture content of wheat and flour. *Cereal Chemistry* 6:379–395.
32. Hubbard, J. E., F. R. Earle, and F. R. Senti. 1957. Moisture relations in wheat and corn. *Cereal Chemistry* 34:422–433.
33. Schroeder, H. W., and J. W. Sorenson, Jr. 1961. Mold development of rough rice as affected by aeration during storage. *Rice Journal* 64:6, 8–10, 12, 21–23.

34. Zeleny, L. 1960. A survey of methods and apparatus for moisture measurement in the grain industry. *Cereal Science Today* 5:130–136.

Chapter 4. Heating and Respiration

35. Bailey, C. H. 1940. Respiration of cereal grains and flaxseed. *Plant Physiology* 15:257–274.
36. Bailey, C. H., and A. J. Gurjar. 1918. Respiration of stored wheat. *Journal of Agricultural Research* 12:685–713.
37. Carter, E. P. 1950. *Role of Fungi in the Heating of Moist Wheat.* USDA Circular 838.
38. Christensen, C. M. 1964. Effect of moisture content and length of storage period upon germination percentage of seeds of corn, wheat, and barley free of storage fungi. *Phytopathology* 54:1464–1466.
39. Cohn, F. 1890. Ueber Wärme Erzeugung durch Schimmelpilze und Bakterien. *Jahresberichte Schles. Gesellschaft* (Breslau) 68:23–29.
40. Coleman, D. A., and H. C. Fellows. 1925. Hygroscopic moisture of cereal grains and flaxseed exposed to atmospheres of different relative humidity. *Cereal Chemistry* 2:275–287.
41. Darsie, M. L., C. Elliott, and G. J. Peirce. 1914. A study of the germinating power of seeds. *Botanical Gazette* 58:101–136.
42. Gilman, J. C., and D. H. Barron. 1930. Effect of molds on temperature of stored grain. *Plant Physiology* 5:565–573.
43. Hummel, B. C. W., L. S. Cuendet, C. M. Christensen, and W. F. Geddes. 1954. Grain Storage Studies 13: Comparative changes in respiration, viability, and chemical composition of mold-free and mold-contaminated wheat upon storage. *Cereal Chemistry* 31:143–150.
44. Larmour, R. K., J. S. Clayton, and C. L. Wrenshall. 1935. A study of the respiration of damp wheat. *Canadian Journal of Research* 12:627–645.
45. Malowan, J. 1921. Some observations on the heating of cottonseed. *Cotton Oil Press* 4:47–49.
46. Milner, M., C. M. Christensen, and W. F. Geddes. 1947. Grain Storage Studies 7: Influence of certain mold inhibitors on respiration of moist wheat. *Cereal Chemistry* 24:507–517.
47. Milner, M., and W. F. Geddes. 1945. Grain Storage Studies 2: The effect of aeration, temperature, and time of the respiration of soybeans containing excessive moisture. *Cereal Chemistry* 22:484–501.
48. Milner, M., and W. F. Geddes. 1946. Grain Storage Studies 3: The relation between moisture content, mold growth, and respiration of soybeans. *Cereal Chemistry* 23:225–247.
49. Moore, M. B., and C. R. Olien. 1952. Mercury bichloride solution disinfectant for cereal seeds. *Phytopathology* 42:471. (Abstract.)
50. Olien, C. R., and M. B. Moore. 1954. Certain mercurial seed treatments do not kill fungi on seed wheat prior to planting. *Phytopathology* 44:500. (Abstract.)
51. Ramstad, P. E., and W. F. Geddes. 1943. *The Respiration and Storage Behavior of Soybeans.* Minnesota Agricultural Experiment Station Technical Bulletin 156.

Chapter 5. Germinability, Discoloration, and Fat Acidity Values

See also references 20, 21, 26, 38, 45, 48
52. Christensen, C. M. 1955. Grain Storage Studies 21: Viability and moldiness of commercial wheat in relation to the incidence of germ damage. *Cereal Chemistry* 32:507–518.

Grain Storage

53. Christensen, C. M., and P. Linko. 1963. Moisture content of hard red winter wheat as determined by meters and by oven drying, and influence of small differences in moisture content upon subsequent deterioration of the grain in storage. *Cereal Chemistry* 40:129–137.
54. Christensen, C. M., and L. C. López F. 1963. Pathology of stored seeds. *Proceedings of the International Seed Testing Association* 28:701–711.
55. Christensen, C. M., J. H. Olafson, and W. F. Geddes. 1949. Grain Storage Studies 8: Relation of molds in moist stored cottonseed to increased production of carbon dioxide, fatty acids, and heat. *Cereal Chemistry* 26:109–128.
56. Fields, R. W., and T. H. King. 1962. Influence of storage fungi on the deterioration of stored pea seed. *Phytopathology* 52:336–339.
57. Goodman, J. J., and C. M. Christensen. 1952. Grain Storage Studies 11: Lipolytic activity of fungi isolated from stored corn. *Cereal Chemistry* 29:299–308.
58. López, L. C., and C. M. Christensen. 1967. Effect of moisture content and temperature on invasion of stored corn by *Aspergillus flavus*. *Phytopathology* 57:588–590.
59. Milner, M., C. M. Christensen, and W. F. Geddes. 1947. Grain Storage Studies 6: Wheat respiration in relation to moisture content, mold growth, chemical deterioration, and heating. *Cereal Chemistry* 24:182–199.
60. Nagel, C. M., and G. Semeniuk. 1947. Some mold-induced changes in shelled corn. *Plant Physiology* 22:20–23.
61. Papavizas, G. C., and C. M. Christensen. 1960. Grain Storage Studies 29: Effect of invasion by individual species and mixtures of species of *Aspergillus* upon germination and development of discolored germs in wheat. *Cereal Chemistry* 37:197–203.

Chapter 6. Mycotoxins and Grain Quality

See also reference 58
62. Brook, P. J., and E. P. White. 1966. Fungus toxins affecting mammals. *Annual Review of Phytopathology* 4:171–194.
63. Burnside, J. E., W. L. Sippel, J. Forgacs, W. T. Carll, M. B. Atwood, and E. R. Doll. 1957. A disease of swine and cattle caused by eating moldy corn, II: Experimental production with pure cultures of molds. *American Journal of Veterinary Research* 18:817–824.
64. Christensen, C. M., H. A. Fanse, G. H. Nelson, Fern Bates, and C. J. Mirocha. 1967. Microflora of black and red pepper. *Applied Microbiology* 15:622–626.
65. Christensen, C. M., G. H. Nelson, and C. J. Mirocha. 1965. Effect on white rat uterus of a toxic substance isolated from *Fusarium*. *Applied Microbiology* 13:653–659.
66. Forgacs, J. 1962. Mycotoxicoses, the neglected diseases. *Feedstuffs* 34:124–134.
67. Graham, R. 1936. Cornstalk disease investigations. *Veterinary Medicine* 31:46–50.
68. Joffe, A. Z. 1965. Toxin production by cereal fungi causing toxic alimentary aleukia in man. In *Mycotoxins in Foodstuffs*, ed. G. N. Wogan, pp. 77–85. M.I.T. Press, 50 Ames Street, Cambridge, Mass. 02142.
69. McNutt, S. H., P. Purwin, and C. Murray. 1928. Vulvovaginitis in swine. *Journal of American Veterinary Medicine Association* 73:484–492.
70. Michener, J. C. 1882. Cerebrospinal meningitis-fungosus toxicum paralyticus. *American Veterinary Revue* 6:345–347.
71. Nelson, G. H., C. M. Christensen, and C. J. Mirocha. 1965. A veterinarian

looks at moldy corn. In *Proceedings of the 20th Annual Hybrid Corn Industry-Research Conference*, pp. 86–91.

72. Purchase, I. F. H., and J. J. Theron. 1967. Research on mycotoxins in South Africa. *International Pathology* 8:3–7.
73. Sippel, W. L., J. E. Burnside, and B. F. Atwood. 1953. A disease of swine and cattle caused by eating molded corn. In *Proceedings of the Veterinary Association*, pp. 174–181.
74. Taber, Ruth A., and H. W. Schroeder. 1967. Aflatoxin-producing potentia! of isolates of the *Aspergillus flavus-oryzae* group from peanuts (*Arachis hypogaea*). *Applied Microbiology* 15:140–144.
75. Theron, J. J., R. J. van der Merwe, N. Liebenberg, H. J. B. Joubert, and W. Nel. 1966. Acute liver injury in ducklings and rats as a result of ochratoxin poisoning. *Journal of Pathology and Bacteriology* 91:521–529.
76. Wilson, B. J. 1965. Toxic substances formed by filamentous fungi growing on feedstuffs. In *Mycotoxins in Foodstuffs*, ed. G. N. Wogan, pp. 147–149. M.I.T. Press, 50 Ames Street, Cambridge, Mass. 02142.
77. Wogan, G. N., ed. 1965. *Mycotoxins in Foodstuffs*. M.I.T. Press, 50 Ames Street, Cambridge, Mass. 02142.

Chapter 7. Evaluation of Condition and Storability

See also references 2, 22

78. Besley, H. J., and G. H. Baston. 1914. *Acidity as a Factor in Determining the Degree of Soundness in Corn*. USDA Bulletin 102.
79. Christensen, C. M., and H. H. Kaufmann. 1968. *Maintenance of Quality in Stored Grains and Seeds*. University of Minnesota Extension Folder 226, revised.
80. Del Prado, F. A., and C. M. Christensen. 1952. Grain Storage Studies 12: The fungus flora of stored rice seed. *Cereal Chemistry* 29:456–462.
81. Koehler, B. 1957. *Pericarp Injuries in Seed Corn: Prevalence in Dent Corn and Relation to Seedling Blights*. Illinois University Agricultural Experiment Station Bulletin 617.
82. Oxley, T. A. 1948. *The Scientific Principles of Grain Storage*. Northern Publishing Co., Liverpool, England.
83. Sorger-Domenigg, H., L. S. Cuendet, C. M. Christensen, and W. F. Geddes. 1955. Grain Storage Studies 17: Effect of mold growth during temporary exposure of wheat to high moisture contents upon the development of germ damage and other indices of deterioration during subsequent storage. *Cereal Chemistry* 32:270–285.
84. Welty, R. E., S. A. Qasem, and C. M. Christensen. 1963. Tests of corn stored four years in a commercial bin. *Cereal Chemistry* 40:277–282.
85. Zeleny, L., and D. A. Coleman. 1938. Acidity in cereals and cereal products, its determination and significance. *Cereal Chemistry* 15:580–595.
86. Zeleny, L., and D. A. Coleman. 1939. *The Chemical Determination of Soundness in Corn*. USDA Technical Bulletin 644.

Chapter 8. Drying, Aeration, and Refrigeration

See also references 1, 22

87. Christensen, C. M. 1967. Some changes in No. 2 corn stored two years at moisture contents of 14.5 and 15.2% and temperatures of 12, 20, and 25 C. *Cereal Chemistry* 44:95–99.
88. Hukill, W. V. 1954. Drying of grain. In *Storage of Cereal Grains and Their Products*, Chapter 9. American Association of Cereal Chemists, 1955 University Avenue, St. Paul, Minn. 55104.

89. Johnson, H. K. 1957. Cooling stored grain by aeration. *Agricultural Engineering* 38:238–241, 244–246.
90. Kaufmann, H. H. 1963. Advances in grain storage. *Cargill Crop Bulletin* 38(12):14–16.
91. Kelly, C. F. 1940. *Methods of Ventilating Wheat in Farm Storage.* USDA Circular 544.
92. Paterson, H. 1967. Chilled grain storage. *Farm Mechanization and Buildings,* August, pp. 19–20.
93. Shove, G. C. 1967. Status of chilled corn storage. In *Proceedings of Iowa Elevator Operators Grain Conditioning Conference,* Iowa State University of Science and Technology, pp. 12–19.

Chapter 9. Insects, Mites, and Rodents

See also references 4, 13, 16, 23
94. Burgess, E. C. 1966. The federal-state war on pests. In USDA *Yearbook of Agriculture,* pp. 258–265. U.S. Government Printing Office, Washington, D.C. 20402.
95. Cotton, R. T. 1950. *Insect Pests of Stored Grain and Grain Products.* Burgess Publishing Co., 426 South 6th Street, Minneapolis, Minn. 55415.
96. Cotton, R. T., and H. E. Gray. 1948. *Preservation of Grains in Storage,* FAO Agricultural Study No. 2, pp. 35–71. Food and Agriculture Organization of the United Nations, Washington, D.C.
97. Cotton, R. T., H. H. Walkden, G. D. White, and D. A. Wilbur. 1953. *Causes of Outbreaks of Stored-Grain Insects.* Kansas Experiment Station Bulletin 359.
98. Crucillo, C. V., F. Q. Otanes, and J. L. Morales. 1954. What we're doing to control field rats in Cotabato, Philippine Islands. *Pest Control,* October, pp. 9–16, 72–76; November, pp. 10–16.
99. Food and Agriculture Organization of the United Nations. 1948. *Preservation of Grains in Storage.* FAO Agricultural Study No. 2. Washington, D.C.
100. Henderson, L. S., and C. M. Christensen. 1961. Postharvest control of insects and fungi. In USDA *Yearbook of Agriculture,* pp. 348–356. U.S. Government Printing Office, Washington, D.C. 20402.
101. Hughes, A. M. 1961. *The Mites of Stored Food.* Ministry of Agriculture, Fisheries and Food Technical Bulletin No. 9. Her Majesty's Stationery Office, London, England.
102. Malhotra, S. P. 1967. Hindu rituals and beliefs toward animal life. *Bulletin of Grain Technology* 5:135–136.
103. Sinha, R. N. 1964. Mites of stored grain in western Canada. Ecology and survey. *Proceedings of the Entomological Society of Manitoba* 20:19–33.
104. Slabodnik, M. 1962. Fumigating ships in harbor. *Pest Control,* July, pp. 30, 32, 34, 38, 40.
105. *Thieves of Stored Grain.* 1948. Food and Agriculture Organization of the United Nations, Washington, D.C.
106. *Transactions of the British Entomological Society of London.* 1870. P. xv.
107. USDA Agricultural Research Service. 1955. *A Situation Report, the Khapra Beetle.*
108. USDA Agricultural Research Service, Plant Pest Control Division. 1965. *Cooperative Economic Insect Report* 15(17):397.
109. USDA Agricultural Research Service, Plant Pest Control Division. 1966. *Cooperative Economic Insect Report* 16(13):269.
110. USDA Agricultural Research Service, Plant Pest Control Division. 1966. *Cooperative Economic Insect Report* 16(15):319.
111. Williams, N. H. 1967. The importance and simplicity of a sanitation program. *Cargill Crop Bulletin,* 42(3):13–14.

Index

INDEX

Aeration, 38, 44, 114–121
Aflatoxins, 77–82, Plate 10
Africa, 3, 14, 78, 82, 88–89, 128
Agrawal, N. S., 34
Alsberg, C. L., 8, 95
Alternaria, 18, 22, 24, 89, 90
American Journal of Veterinary Research, 77
Arizona, 129, 130
Aspergillus, 21, 24, 27, 28, 71, 72, 89, 90
Aspergillus amstelodami, 20, 26, 27, 74
Aspergillus candidus, 22, 24, 27, 28, 34, 60, 71, 74
Aspergillus flavus, 22, 28, 34, 63, 64, 66, 74, 76, 77, 79–82, Plate 10
Aspergillus glaucus, 20, 22, 26, 27, 28, 63, 64, 66, 71, 100
Aspergillus halophilicus, 25
Aspergillus ochraceus, 82
Aspergillus repens, 20, 24, 26, 27, 34
Aspergillus restrictus, 20, 22, 24, 25, 26, 27, 30, 34, 75
Aspergillus ruber, 20, 24, 26, 27

Bailey, C. H., 60
Bailey, J. E., 16, 110–111
Barley, 3, 17, 18, 19, 20, 21, 57, 66, 68, 69, 116, 135, Plate 1
Barron, D. H., 57
Baston, G. H., 95
Beans, 3, 20, 21, 67, 99
Besley, H. J., 95
Black, O. F., 8, 95
Boerner, E. G., 8
Brazil, 9, 77, 78
Brook, P. J., 80, 82, 87

Cairo, Ill., 51

California, 128, 129
Canada, 14, 21, 34, 130, 134, 135–136
Cargill, Inc., 49, 72, 101, 107, 118, 120, 133
Cereal Chemistry, 55
Chicago, Ill., 40, 69, 70, 107
Chicks, 83, 87, Plate 10
Christensen, C. M., 20, 21, 22, 23, 28, 29, 31, 43, 65, 66, 67, 69, 71, 72, 74, 75, 82, 83, 101, 102–104, 122, 134
Cladosporium, 18, 90
Coleman, D. A., 55, 95, 96
Corn: production of, 3; nature of losses, 5–10; standards for, 14, 15; field fungi and, 17–18; storage fungi and, 20, 21, 23, Plates 1, 5; moisture content in, 25, 27, 28, 38–39, 48–49, 108; length of storage, 27–30; temperature and, 28, 29, 112, 122; degree of invasion, 31, 32–33; germinability, 65–66, 67, 69, 112; discoloration, 66, 71; fat acidity, 74–75, 95; toxic, 76–77, 79, 82, 83, 86, 87, 89, 90, 91; storability, 102–107, 108; drying, 109–114 *passim*; aeration, 115–118; seed corn cared for, 135; and *passim*
Cotabato, 137–138
Cotton, R. T., 124, 132, 134
Cottonseeds, 60, 73–74
Culturing to detect fungus invasion, 23–24, 99–101

Darsie, M. L., 56, 57
Del Prado, F. A., 101
Dermestid beetles, 126, 128–130
Discoloration, caused by storage fungi, 4, 69–72
Drescher, R. F., 43
Drying, 109–114

151